Springer Series on Epidemiology and Public Health

Series Editors
Wolfgang Ahrens
Iris Pigeot

For further volumes:
http://www.springer.com/series/7251

Suhail A.R. Doi • Gail M. Williams
Editors

Methods of Clinical Epidemiology

Editors
Suhail A.R. Doi
Clinical Epidemiology Unit
School of Population Health
University of Queensland
Herston, Queensland
Australia

Gail M. Williams
Department of Epidemiology & Biostatistics
School of Population Health
University of Queensland
Herston, Queensland
Australia

ISSN 1869-7933 ISSN 1869-7941 (electronic)
ISBN 978-3-642-37130-1 ISBN 978-3-642-37131-8 (eBook)
DOI 10.1007/978-3-642-37131-8
Springer Heidelberg New York Dordrecht London

Library of Congress Control Number: 2013940612

Printed on acid-free paper

Springer is part of Springer Science+Business Media (www.springer.com)

Preface

This book was written to fill the gap that exists in the methods of epidemiology of interest to clinical researchers. It will enable a reader who is currently undertaking research to get key information regarding methodology. It will also help health care personnel from all fields (doctors, nurses, allied health, dentists, pharmacists, etc.) to obtain an effective understanding of methodology useful to research in their field as we cover the unique methods not covered properly in current research methods texts. The classic theoretical focus is avoided because we believe that research must be based on understanding guided by the reader's knowledge of the methodology.

Part I begins by introducing readers to the methods used in clinical agreement studies. It is written to suit beginners but without turning off intermediate users. Qualitative and quantitative agreement are presented, and this section explains how we can utilize these methods and their strengths and weaknesses. Part II shows readers how they can interpret and conceptualize diagnostic test methodologies and ends with an introduction to diagnostic meta-analyses. Part III takes the reader through methods of regression for the binomial family as well as survival analysis and Cox regression. Here, the focus is on methods of use to clinical researchers. These methods have different names and multiple interpretations, which are explained. It is important to know what associations you are interested in and know what data are available and in what form they can be used. An in-depth discussion of the use of these methods is presented with a view to giving the reader a clear understanding of their utility and interpretation. Part IV deals with systematic reviews and meta-analyses. A step-by-step approach is used to guide readers through the key principles that must be understood before undertaking a meta-analysis, with particular emphasis on newer methods for bias adjustment in meta-analysis, an area in which we have considerable expertise.

We thank Lorna O'Brien from authorserv.com for her dedicated help with the editing of this book and Federica Corradi Dell'Acqua, Editorial Manager for Biomathematics & Statistics at Springer for continuous advice throughout the publication process. Finally, we realize that this first edition may include inconsistencies and mistakes, and we welcome any suggestion from readers to improve its content.

30 June 2012
Brisbane

Suhail A.R. Doi
Gail M. Williams

Acknowledgments

Every effort has been made to trace rights holders, but if any have been inadvertently overlooked the publishers would be pleased to make the necessary arrangements at the first opportunity.

List of Abbreviations

ADA	Adenosine deaminase activity
AGME	Accreditation Council for Graduate Medical Education
ANA	Antinuclear antibodies
ANOVA	Analysis of variance
AUC	Area under the curve
BAK	Bias-adjusted kappa
BF	Body fat
BMI	Body mass index
BP	Blood pressure
CADTH	Canadian Agency for Drugs and Technology in Health
CASP	Critical Appraisal Skills Programme
CEBM	Centre for Evidence Based Medicine
CF	Correction factor
CHF	Congestive heart failure
CI	Confidence interval
CL	Confidence limits
CT	Computed tomography
CV	Coefficient of variation
DOR	Diagnostic odds ratio
DXA	Dual-energy X-ray absorptiometry
ECT	Electroconvulsive therapy
ELISA	Enzyme-linked immunosorbent assay
EPHPP	Effective Public Health Practice Project
ES	Effect size
ESR	Erythrocyte sedimentation rate
ESS	Effective sample size
FN	False-negative
FP	False-positive
FPR	False-positive rate

GLM	Generalized Linear Model
HIV	Human immunodeficiency virus
HSROC	Hierarchical summary receiver operator characteristic
HTA	Health technology assessment
ICC	Intraclass correlation coefficient
ICC	Intraclass correlation coefficient
IID	Independent and identically distributed
ITT	Intention-to-treat
LAG	Lymphangiography
LCL	Lower confidence limit
LR	Likelihood ratio
MERGE	Method for Evaluating Research and Guideline Evidence
MeSH	Medical Subject Heading
MI	Myocardial infarction
MRI	Magnetic resonance imaging
MSE	Mean squared error
NLM	National Library of Medicine
NOS	Newcastle-Ottawa Scale
NPV	Negative predictive value
NSS	Numerical sum score
NTP	Negative test probability
OCLC	Online Computer Library Center
OFIA	Operational financial impact assessment
PABAK	Prevalence-adjusted-bias-adjusted kappa
PICO	Population, intervention or exposure, comparison, outcome
PPV	Positive predictive value
PT	Pertussis toxin
QCR	Qualitative rating on level of components
QE	Quality effect
QOR	Qualitative overall rating
QUADAS	Quality Assessment of Diagnostic Accuracy Studies
RCT	Randomized controlled trial
RE	Random effect
REVC	Random effects variance component
RLR	Ratio of the likelihood ratio
ROC	Receiver operating characteristic
ROM	Range of motion
RR	Relative risks
SD	Standard deviation
SE	Standard error
SEM	Standard error of the measurement
SLE	Systemic lupus erythematosus
SMD	Standardized mean difference
TcB	Transcutaneous bilirubin

TN	True-negative
TP	True-positive
TPR	True-positive rate
TSB	Total serum bilirubin
UCL	Upper confidence limit
WHO	World Health Organization

Contents

Part III Modeling Binary and Time-to-Event Outcomes

Part IV Systematic Reviews and Meta-analysis

Contributors

Víctor Abraira Clinical Biostatistics Unit, Hospital Universitario Ramón y Cajal, CIBER en Epidemiología y Salud Pública (CIBERESP) and Instituto Ramón y Cajal de Investigación Sanitaria (IRYCIS), Madrid, Spain

Cristian Baicus Clinical Epidemiology Unit, Bucharest, and Associate Professor of Internal Medicine and Carol Davila University of Medicine and Pharmacy, Bucharest, Romania

Jan J. Barendregt University of Queensland, School of Population Health, Brisbane, Australia

Justin Clark University of Queensland, Brisbane, Australia

Suhail A.R. Doi University of Queensland, School of Population Health, Brisbane, Australia and Princess Alexandra Hospital, Brisbane, Australia

Maren Dreier Hannover Medical School, Institute for Epidemiology, Social Medicine and Health Systems Research, Hannover, Germany

Rosmin Esmail Knowledge Translation, Research Portfolio, Alberta Health Services, Calgary, AB, Canada

Abhaya Indrayan Department of Biostatistics and Medical Informatics, University College of Medical Sciences, Delhi, India

Rajeev Kumar Malhotra Department of Biostatistics and Medical Informatics, University College of Medical Sciences, New Delhi, India

Adedayo A. Onitilo Marshfield Clinic - Weston Center, Weston, USA and University of Wisconsin Medical School, Madison, USA

María Nieves Plana Clinical Biostatistics Unit, Hospital Universitario Ramón y Cajal, CIBER en Epidemiología y Salud Pública (CIBERESP) and Instituto Ramón y Cajal de Investigación Sanitaria (IRYCIS), Madrid, Spain

Orit Shechtman Department of Occupational Therapy, College of Public Health and Health Professions, University of Florida, Gainesville, FL, USA

Sophie Vanbelle Department of Methodology and Statistics, School of Public Health and Primary Care, Maastricht University, Maastricht, The Netherlands

Robert Ware University of Queensland, School of Population Health, Brisbane, Australia

Gail M. Williams University of Queensland, School of Population Health, Brisbane, Australia

Javier Zamora Clinical Biostatistics Unit, Hospital Universitario Ramón y Cajal, CIBER en Epidemiología y Salud Pública (CIBERESP) and Instituto Ramón y Cajal de Investigación Sanitaria (IRYCIS), Madrid, Spain

Part I
Clinical Agreement

Chapter 1
Clinical Agreement in Qualitative Measurements

The Kappa Coefficient in Clinical Research

Sophie Vanbelle

Abstract Agreement between raters on a categorical scale is not only a subject of scientific research but also a problem frequently encountered in practice. For example, in psychiatry, the mental illness of a subject may be judged as light, moderate or severe. Inter- and intra-rater agreement is a prerequisite for the scale to be implemented in routine use. Agreement studies are therefore crucial in health, medicine and life sciences. They provide information about the amount of error inherent to any diagnostic, score or measurement (e.g. disease diagnostic or implementation quality of health promotion interventions). The kappa-like coefficients (intraclass kappa, Cohen's kappa and weighted kappa), usually used to assess agreement between or within raters on a categorical scale, are reviewed in this chapter with emphasis on the interpretation and the properties of these coefficients.

Introduction

The problem of rater agreement on a categorical scale originally emerged in human sciences, where measurements are traditionally made on a nominal or ordinal scale rather than on a continuum. For example, in psychiatry, the mental illness of a subject may be judged as light, moderate or severe. Clearly two psychiatrists assessing the mental state of a series of patients do not necessarily give the same grading for each patient but we would expect some agreement between them (inter-rater agreement). In the same way, we could observe some variation in the assessment of the same patients by a psychiatrist on two occasions (intra-rater agreement). Agreement studies, therefore, became crucial in health, medicine and

S. Vanbelle (✉)
Department of Methodology and Statistics, School of Public Health and Primary Care, Maastricht University, Maastricht, The Netherlands
e-mail: Sophie.vanbelle@Maastrichtuniversity.nl

S.A.R. Doi and G.M. Williams (eds.), *Methods of Clinical Epidemiology*, Springer Series on Epidemiology and Public Health,
DOI 10.1007/978-3-642-37131-8_1,

life sciences. They provide information about the amount of error inherent to any diagnosis, score or measurement.

Agreement has to be distinguished from the concept of reliability. When elements (objects, subjects, patients, items) are evaluated by two raters (observers, judges, methods), agreement refers to the degree of closeness between the two assessments within an element (i.e. classification of each element in the same category by the two raters). By contrast, reliability refers to the degree of differentiation between the elements (i.e. the two raters give the same relative ordering of the elements). Good reliability is essential when the purpose is to assess the correlation with other measures (e.g. severity of the mental illness and autonomy) because of the well-known attenuation effect. Good agreement is, on the other hand, imperative for clinical decision making (e.g. prescribing a treatment for a specific patient based on the seriousness of the mental illness). Two kinds of agreement are usually distinguished. Inter-rater agreement refers to a sample of elements assessed with the same instrument by different raters; the term intra-rater agreement is used when a sample of elements is assessed on two occasions by the same rater using the same instrument.

Several coefficients for quantifying the agreement between two raters on a categorical scale have been introduced over the years. Cohen's (1960) kappa coefficient is the most salient and the most widely used in the scientific literature. Cohen (1968) extended the kappa coefficient to weighted kappa coefficients to allow for some more important disagreements than others (e.g. disagreements between the categories light and severe may be viewed as more important than between light and moderate). Kraemer (1979) defined the intraclass kappa coefficient by assuming that the two raters have the same marginal probability distribution. All these coefficients belong to the kappa-like family and possess the same characteristic: they account for the occurrence of agreement due to chance only.

An example used through this chapter to illustrate the use and the computation of the various kappa coefficients is presented in the next section. The third section focuses on binary scales. Kappa coefficients are introduced for nominal and ordinal scales in the fourth and fifth sections, respectively. Then, before drawing conclusions, the distinction between the concepts of agreement and association is illustrated on an example.

Example

Cervical ectopy, defined as the presence of endocervical-type columnar epithelium on the portio surface of the cervix, has been identified as a possible risk factor for heterosexual transmission of human immunodeficiency virus. Methods for measuring the cervical ectopy size with precision are therefore needed. Gilmour et al. (1997) conducted a study to compare the agreement obtained between medical raters by direct visual assessment and a new computerized planimetry method. Photographs of the cervix of 85 women without cervical disease were assessed for cervical ectopy by three medical raters who used both

Table 1.1 4 × 4 contingency table resulting from the direct visual assessment of cervical ectopy size by two medical raters on 85 women in terms of frequency

	Medical rater 2				
Medical rater 1	Minimal	Moderate	Large	Excessive	Total
Minimal	13	2	0	0	15
Moderate	10	16	3	0	29
Large	3	7	3	0	13
Excessive	1	4	12	11	28
Total	27	29	18	11	85

assessment methods. The response of interest, cervical ectopy size, was an ordinal variable with $K = 4$ categories: (1) minimal, (2) moderate, (3) large and (4) excessive. The classification of the 85 women by two of the three raters via direct visual assessment is summarized in Table 1.1 in terms of frequency. We will determine the agreement between these two raters on each category separately and on the four-point scale.

Binary Scale

The simplest case is to determine the agreement between two raters who have to classify a sample of N elements (subjects, patients or objects) into two exhaustive and mutually exclusive categories (e.g. diseased/non-diseased). For example, women can be classified as having (1) or not having (0) an excessive ectopy size. The observations made by the two raters can be summarized in a 2 × 2 contingency table (Table 1.2), where n_{jk} is the number of elements classified in category j by rater 1 and category k by rater 2, $n_{j.}$ the number of elements classified in category j by rater 1 and $n_{.k}$ the number of elements classified in category k by rater 2. By dividing these numbers by the total number of observations N, the corresponding proportions p_{jk}, $p_{j.}$, $p_{.k}$ are obtained $(j, k = 1, 2)$. The proportions $p_{1.}$ and $p_{2.}$ determine the marginal distribution of rater 1 and $p_{.1}$ and $p_{.2}$ the marginal distribution of rater 2. The marginal distribution refers to the distribution of the classification of one rater, irrespective of the other rater's classification.

Cohen's Kappa Coefficient

Intuitively, it seems obvious to use the sum of the diagonal elements in Table 1.2 to quantify the level of agreement between the two raters. It is the proportion of elements classified in the same category by the two raters. This simplest agreement index, usually denoted by p_o, is called the *observed proportion of agreement*,

$$p_o = \frac{n_{11} + n_{22}}{N} = p_{11} + p_{22}.$$

Table 1.2 2×2 contingency table corresponding to the classification of N elements on a binary scale by two raters in terms of frequency (left) and proportion (right)

	Rater 2				Rater 2		
Rater 1	1	0	Total	Rater 1	1	0	Total
1	n_{11}	n_{12}	$n_{1.}$	1	p_{11}	p_{12}	$p_{1.}$
0	n_{21}	n_{22}	$n_{2.}$	0	p_{21}	p_{22}	$p_{2.}$
Total	$n_{.1}$	$n_{.2}$	N	Total	$p_{.1}$	$p_{.2}$	1

However, this coefficient does not account for the fact that a number of agreements between the two raters can occur purely by chance. If the two raters randomly assign the elements on a binary scale (e.g. based on the results of a tossed coin), the proportion of agreement between them is only attributable to chance. Therefore, Cohen (1960) introduced the *proportion of agreement expected by chance* as

$$p_{\mathrm{e}} = \frac{n_{1.}n_{.1} + n_{2.}n_{.2}}{N^2} = p_{1.}p_{.1} + p_{2.}p_{.2}$$

It is the proportion of agreement expected if rater 1 classifies the elements randomly with a marginal distribution $(p_{.1}, p_{.2})$ and rater 2 with a marginal distribution $(p_{1.}, p_{2.})$. Cohen corrected the observed proportion of agreement for the proportion of agreement expected by chance and scaled the result to obtain a value 1 when agreement is perfect (all observations fall in the diagonal cells of the contingency table), a value 0 when agreement is only to be expected by chance and negative values when the observed proportion of agreement is lower than the proportion of agreement expected by chance (with a minimum value of -1). Specifically, Cohen's kappa coefficient is written as

$$\hat{\kappa} = \frac{p_{\mathrm{o}} - p_{\mathrm{e}}}{1 - p_{\mathrm{e}}}. \tag{1.1}$$

Cohen's kappa coefficient is more often used to quantify inter-rater agreement than intra-rater agreement because it does not penalize the level of agreement for differences in the marginal distribution of the two raters (i.e. when $p_{1.} \neq p_{.1}$). Different marginal distributions are expected in the presence of two raters with different work experience, background or using different methods.

Intraclass Kappa Coefficient

The intraclass kappa coefficient was derived by analogy to the intraclass correlation coefficient for continuous outcomes and is based on the common correlation model. This model assumes that the classifications made by the two raters are interchangeable. In other words, the two raters are supposed to have the same marginal probability distribution (i.e. the probability of classifying an element in category

1 is the same for the two raters). This is typical of a test–retest situation where there is no reason for the marginal probabilities to change between the two measurement occasions. The resulting index is algebraically equivalent to Scott's index of agreement and can be viewed as a special case of Cohen's kappa coefficient.

The *observed proportion of agreement* is the same as in the case of Cohen's kappa coefficient

$$p_{oI} = \frac{n_{11} + n_{22}}{N} = p_{11} + p_{22} = p_o$$

but the proportion of agreement expected by chance is determined by

$$p_{eI} = \bar{p}_1^2 + (1 - \bar{p}_1)^2$$

where $\bar{p}_1$ estimates the probability, common to the two raters, of classifying an element in category 1, namely $\bar{p}_1 = (p_{1.} + p_{.1})/2$. The intraclass kappa coefficient is then defined by

$$\hat{\kappa}_I = \frac{p_{oI} - p_{eI}}{1 - p_{eI}}. \tag{1.2}$$

Interpretation

Two main criticisms on kappa coefficients were formulated in the literature. First, like correlation coefficients, kappa coefficients vary between −1 and +1 and have no clear interpretation, except for 0 and 1 values. Landis and Koch (1977) proposed qualifying the strength of agreement according to the values taken by the kappa coefficient. This classification is widely used but should be avoided because its construction is totally arbitrary and the value of kappa coefficients depends on the prevalence of the trait studied. It is preferable to consider a confidence interval to appreciate the value of a kappa estimate; often only the lower bound is of interest. Several methods were derived to estimate the sampling variability of kappa-like agreement coefficients. Most of the statistical packages (e.g. SAS, SPSS, STATA, R) report the sample variance given by the delta method. The formula is given in the Appendix for the general case of more than two categories.

Second, several authors pointed out that kappa coefficients are dependent on the prevalence of the trait under study, which indicates a serious limitation when comparing values of kappa coefficients among studies with varying prevalence. More precisely, Thompson and Walter (1988) demonstrated that kappa coefficients can be written as a function of the true prevalence of the trait, as well as the sensitivity and the specificity of each rater classification. This dependence can lead to surprising results when a high observed proportion of agreement is associated with a low kappa value.

Some alternatives to the classic Cohen's kappa coefficient have been proposed to cope with this problem. For example, the bias-adjusted kappa (BAK) allows adjustment of Cohen's kappa coefficient for rater bias (i.e. differences in the marginal probability distribution of the two raters). The BAK coefficient turns out to be equivalent to the intraclass kappa coefficient $\hat{\kappa}_I$ defined in Eq. 1.2. Furthermore, a prevalence-adjusted-bias-adjusted kappa (PABAK), which is nothing more than a linear transformation of the observed proportion of agreement ($\text{PABAK} = 2p_o - 1$), was suggested by Byrt et al. (1993).

Therefore, despite its drawbacks, Cohen's kappa coefficient remains popular to assess agreement in the absence of a gold standard. However, it should be kept in mind that Cohen's kappa coefficient mixes two sources of disagreement among raters: disagreement due to bias among raters (i.e. different probabilities to classify elements in category 1 for the two raters) and disagreement that occurs because the raters evaluate the elements differently (i.e. rank order the elements differently). Rater bias can be studied by comparing values of the kappa coefficient and the intraclass kappa coefficient. Cohen' kappa coefficient is always larger than the intraclass kappa coefficient because it does not penalize for rater bias, equivalence being reached when there is no rater bias ($n_{12} = n_{21}$). Therefore, the larger the difference between the two coefficients, the larger the rater bias. On the other hand, a difference between the intraclass kappa coefficient (BAK) and PABAK indicates that the marginal probability distributions of the raters depart from the uniform distribution ($\bar{p}_1 = \bar{p}_2 = 0.5$).

Example

Consider the cervical ectopy data given in Table 1.1, where two medical raters classify the cervical ectopy size of 85 women. To determine the agreement on each category separately, 2×2 tables were constructed by isolating one category and collapsing all the other categories together (Table 1.3).

When considering the category minimal against all other categories, the observed proportion of agreement is equal to

$$p_o = \frac{13 + 56}{85} = 0.81.$$

This means that the two medical raters classify 81 % of the women in the same category, that is, they agree on 81 % of the women. The proportion of agreement expected by chance is equal to

$$p_e = \frac{27 \times 15 + 58 \times 70}{85^2} = 0.62.$$

Table 1.3 Contingency tables obtained from the classification of the ectopy size of 85 women by two medical raters with direct visual assessment when isolating each category of the four-category scale

Category minimal

	Rater 2		
Rater 1	Minimal	Other	Total
Minimal	13	2	15
Other	14	56	70
Total	27	58	85

Category moderate

	Rater 2		
Rater 1	Moderate	Other	Total
Moderate	16	13	29
Other	13	43	56
Total	29	56	85

Category large

	Rater 2		
Rater 1	Large	Other	Total
Large	3	10	13
Other	15	57	72
Total	18	67	85

Category excessive

	Rater 2		
Rater 1	Excessive	Other	Total
Excessive	11	17	28
Other	0	57	57
Total	11	74	85

Therefore, given the marginal distribution of the two medical raters, if they classify the elements randomly, we expect them to agree on 62 % of the women. This leads to a Cohen's kappa coefficient of

$$\hat{\kappa} = \frac{p_o - p_e}{1 - p_e} = \frac{0.81 - 0.62}{1 - 0.62} = 0.51.$$

In a same way, the intraclass kappa coefficient is equal to 0.49. The results obtained for the other categories are summarized in Table 1.4.

It can be observed in Table 1.4 that there is a significant positive agreement on all categories, except on category large (the lower bound of the 95 % confidence interval is negative). More generally, it is seen that the agreement on extreme categories (minimal and excessive) is better than the agreement on the middle categories (moderate and large). This is a well-know phenomenon. When the marginal distributions of the two raters are the same (see category moderate in Table 1.3), we have $\hat{\kappa} = \hat{\kappa}_I$, as expected.

Categorical Scale

Cohen's Kappa and Intraclass Kappa Coefficients

Consider now the situation where two raters have to classify N elements on a categorical scale with more than two ($K > 2$) categories (e.g. cervical ectopy size is rated on a four-category scale). By extension, Cohen (1960) defined the *observed proportion of agreement* and the *proportion of agreement expected by chance* by

Table 1.4 Observed proportion of agreement, proportion of agreement expected by chance, kappa coefficient, standard error and 95 % confidence interval (95 % CI) of the Cohen's kappa coefficient and the intraclass kappa coefficient for each 2 × 2 table given in Table 1.3

	Cohen's kappa				
Category	p_o	p_e	$\hat{\kappa}$	SE($\hat{\kappa}$)	95 % CI
Minimal	0.81	0.62	0.51	0.10	0.31, 0.71
Moderate	0.69	0.55	0.32	0.11	0.11, 0.53
Large	0.71	0.70	0.019	0.11	−0.19, 0.23
Excessive	0.80	0.63	0.47	0.098	0.27, 0.66
	Intraclass kappa				
	p_{oI}	p_{eI}	$\hat{\kappa}_I$	SE($\hat{\kappa}_I$)	95 % CI
Minimal	0.81	0.63	0.49	0.11	0.27, 0.72
Moderate	0.69	0.55	0.32	0.11	0.11, 0.53
Large	0.71	0.70	0.014	0.13	−0.24, 0.27
Excessive	0.80	0.65	0.43	0.11	0.23, 0.64

$$p_o = \sum_{j=1}^{K} \frac{n_{jj}}{N} = \sum_{j=1}^{K} p_{jj} \quad \text{and} \quad p_e = \sum_{j=1}^{K} \frac{n_{j.}n_{.j}}{N^2} = \sum_{j=1}^{K} p_{j.}p_{.j}.$$

This leads to the Cohen's kappa coefficient

$$\hat{\kappa} = \frac{p_o - p_e}{1 - p_e}.$$

In the same way, we have for the intraclass kappa coefficient

$$p_{oI} = \sum_{j=1}^{K} p_{jj}, \quad p_{eI} = \sum_{j=1}^{K} \left(\frac{p_{j.} + p_{.j}}{2}\right)^2 \quad \text{and} \quad \hat{\kappa}_I = \frac{p_{oI} - p_{eI}}{1 - p_{eI}}.$$

Interpretation

It has been proven that Cohen's kappa and the intraclass kappa coefficients computed for a $K \times K$ contingency table are in fact weighted averages of kappa coefficients obtained on 2×2 tables, constructed by isolating a single category $[j]$ from the other categories (see Table 1.3) ($j = 1, \ldots, K$). The overall proportion of observed agreement is in fact the average of the observed proportion of agreement in the 2×2 tables and the same applies for the proportion of agreement expected by chance. More precisely, we have

$$\hat{\kappa} = \frac{\sum_{j=1}^{K} (p_{o[j]} - p_{e[j]})}{\sum_{j=1}^{K} (1 - p_{e[j]})} = \frac{1}{\sum_{j=1}^{K} (1 - p_{e[j]})} \sum_{j=1}^{K} (1 - p_{e[j]}) \hat{\kappa}_{[j]}.$$

Example

In the cervical ectopy example, the proportion of observed agreement and the proportion of agreement expected by chance are respectively equal to

$$p_o = (13 + 16 + 3 + 11)/85 = 0.506$$

and

$$p_e = (27 \times 15 + 29 \times 29 + 18 \times 13 + 11 \times 28)/85^2 = 0.247.$$

Cohen's kappa coefficient is then equal to

$$\hat{\kappa} = \frac{0.506 - 0.247}{1 - 0.247} = 0.34 \quad (95\ \%\ \text{CI}\ 0.21 - 0.48).$$

The average of the observed and expected proportions of agreement in the 2×2 tables (see Table 1.3) are p_o = (0.81 + 0.69 + 0.71 + 0.80)/4 = 0.506 and p_e = (0.62 + 0.55 + 0.70 + 0.63)/4 = 0.247, as expected.

In the same way, the overall intraclass kappa coefficient is equal to $\hat{\kappa}_I = (0.506 - 0.263)/(1 - 0.263) = 0.33$ (95 % CI 0.19–0.47).

Ordinal Scale

Weighted Kappa Coefficients

Some disagreements between two raters can be considered more important than others. For example, on an ordinal scale, disagreements on two extreme categories are generally considered more important than on neighbouring categories. In the cervical ectopy example, discordance between minimal and excessive has more impact than between minimal and moderate. For this reason, in 1968 Cohen introduced the weighted kappa coefficient. Agreement (w_{jk}) or disagreement (v_{jk}) weights are a priori distributed in the K^2 cells of the $K \times K$ contingency table summarizing the classification of the two raters, to reflect the seriousness of disagreement according to the distance between the categories. The weighted kappa coefficient is then defined in terms of agreement weights

$$\hat{\kappa}_w = \frac{p_{ow} - p_{ew}}{1 - p_{ew}} \quad (1.3)$$

with

$$p_{ow} = \sum_{j=1}^{K} \sum_{k=1}^{K} w_{jk} p_{jk} \quad \text{and} \quad p_{ew} = \sum_{j=1}^{K} \sum_{k=1}^{K} w_{jk} p_{j.} p_{.k}$$

(usually $0 \leq w_{jk} \leq 1$ and $w_{jj} = 1$), or in terms of disagreement weights

$$\hat{\kappa}_w = 1 - \frac{q_{ow}}{q_{ew}} \quad (1.4)$$

with

$$q_{ow} = \sum_{j=1}^{K} \sum_{k=1}^{K} v_{jk} p_{jk} \quad \text{and} \quad q_{ew} = \sum_{j=1}^{K} \sum_{k=1}^{K} v_{jk} p_{j.} p_{.k}$$

(usually $0 \leq v_{jk} \leq 1$ and $v_{jj} = 0$).

Although weights can be arbitrarily defined, two agreement weighting schemes defined by Cicchetti and Allison (1971) are commonly used. These are the linear and quadratic weights, given respectively by

$$w_{jk} = 1 - \frac{|j-k|}{K-1} \quad \text{and} \quad w_{jk} = 1 - \left(\frac{|j-k|}{K-1}\right)^2.$$

The disagreement weights $v_{jk} = (j-k)^2$ are also used. Note that Cohen's kappa coefficient is a particular case of the weighted kappa coefficient where $w_{jk} = 1$ when $j = k$ and $w_{jk} = 0$ otherwise.

Interpretation

The use of weighted kappa coefficients was also criticized in the literature, mainly because the weights are generally given a priori and defined arbitrarily. Quadratic weights have received much attention in the literature because of their practical interpretation. For instance, using the disagreement weights $v_{jk} = (j-k)^2$, the weighted kappa coefficient can be interpreted as an intraclass correlation coefficient in a two-way analysis of variance setting (see Fleiss and Cohen (1973); Schuster (2004)).

By contrast, linear weights possess an intuitive interpretation. The $K \times K$ contingency table can be reduced into a 2×2 classification table by grouping the first k categories in one category and the last $K - k$ categories in a second category $(k = 1, \ldots, K - 1)$. The linearly weighted observed and expected agreements are then merely the mean values of the corresponding proportions of all these 2×2 tables. Therefore, similar to Cohen's kappa coefficient, the linearly weighted kappa coefficient is a weighted average of individual kappa coefficients (see Vanbelle and Albert (2009).

The value of the weighted kappa coefficient can vary considerably for different weighting schemes used and henceforth may lead to different conclusions. Clear guidelines for the selection of weights are not yet available in the literature. However, Warrens (2012) tends to favour the use of the linearly weighted kappa because the quadratically weighted kappa is not always sensitive to changes in the diagonal cells of a contingency table.

Example

Consider again the cervical ectopy size example, where women are classified on a four-category Likert scale by two raters (see Table 1.1). The linear and quadratic agreement weights corresponding to the four-category scale are given in Table 1.5. As an illustration, the linear and quadratic weights for the cell (1,2) are equal to $1 - |1 - 2|/(4 - 1) = 0.67$ and $1 - |1 - 2|^2/(4 - 1)^2 = 0.89$, respectively.

To determine the linearly and quadratically weighted kappa coefficient, we have to determine the weighted observed agreement and weighted expected agreement separately. For each of the $K \times K$ cells, we have to multiply the proportion of elements in the cell (p_{jk}) by the corresponding weight (w_{jk}) and then sum these to obtain p_{ow} (Table 1.6). The weighted expected agreement is obtained similarly. For cell (1,2), we have $w_{12}n_{12}/N = 0.67 \times 2/85 = 0.016$ and $0.89 \times 2/85 = 0.021$, respectively.

The linearly weighted kappa coefficient ($\pm$SE) obtained is 0.52 ± 0.060 (95 % CI 0.40–0.64) with $p_{\text{ow}} = 0.80$ and $p_{\text{ew}} = 0.58$. The quadratically weighted kappa coefficient is 0.67 (95 % CI 0.55–0.78) with $p_{\text{ow}} = 0.91$ and $p_{\text{ew}} = 0.72$. In this example, the quadratically weighted kappa coefficient is greater than the linearly weighted kappa coefficient. However, the reverse could happen in other data sets. No clear relationship between the two coefficients has been established in the literature.

Table 1.5 Linear (left) and quadratic (right) weighting schemes for a four-category scale

	Rater 2					Rater 2			
Rater 1	1	2	3	4	Rater 1	1	2	3	4
1	1.00	0.67	0.33	0.00	1	1.00	0.89	0.56	0.00
2	0.67	1.00	0.67	0.33	2	0.89	1.00	0.89	0.56
3	0.33	0.67	1.00	0.67	3	0.56	0.89	1.00	0.89
4	0.00	0.33	0.67	1.00	4	0.00	0.56	0.89	1.00

Table 1.6 Observed agreement in each cell using the linear (left) and quadratic (right) weighting schemes in the cervical ectopy example

	Rater 2					Rater 2			
Rater 1	1	2	3	4	Rater 1	1	2	3	4
1	0.15	0.016	0.00	0.00	1	0.15	0.021	0.00	0.00
2	0.078	0.19	0.024	0.00	2	0.10	0.19	0.031	0.00
3	0.012	0.055	0.035	0.00	3	0.020	0.073	0.035	0.00
4	0.00	0.016	0.094	0.13	4	0.00	0.026	0.13	0.13

Table 1.7 Number of patients classified in one of three diagnostic categories (A, B, C) by two raters

	Rater 2			
Rater 1	A	B	C	Total
A	16	0	24	40
B	20	6	4	30
C	4	14	12	30
Total	40	20	40	100

Agreement Versus Association

A frequent mistake is to use the chi-square test to quantify the agreement between raters. However, the ϕ coefficient, which is equal to Pearson's correlation coefficient for two dichotomous variables, is always larger than Cohen's kappa coefficient, equality holding only when the marginal distribution of the two raters is uniform ($p_{1.} = p_{.1} = 0.5$). The example of Fermanian (1984) illustrates the misuse of association instead of agreement. Let two raters classify independently $N = 100$ patients in three diagnostic categories A, B and C (Table 1.7).

In this example, the value of the chi-square statistic is $\chi^2_{\text{obs}} = 38.7$ with 4 degrees of freedom. Hence, there is a highly significant association between the two ratings ($p < 0.0001$). However, the observed proportion of agreement $p_o = 0.34$ and the proportion of agreement expected by chance $p_e = 0.34$, leading to a Cohen's kappa coefficient of $\hat{\kappa} = (0.34 - 0.34)/(1 - 0.34) = 0$. The other agreement coefficients reviewed in this chapter are also equal to 0. Therefore, despite the existence of a strong association between the two ratings, the agreement between the raters is only to be expected by chance. This example shows that agreement and association measures answer different research questions and should be use in different contexts.

Conclusion

This chapter reviewed the definitions and properties of kappa-like coefficients. These coefficients quantify the amount of agreement beyond chance when two raters classify a series of elements on a categorical scale. Cohen's kappa coefficient and the intraclass kappa coefficient are used when the scale is binary or nominal, whereas weighted kappa coefficients are mainly used for ordinal scales. Despite their controversial properties reviewed in this chapter, kappa coefficients remain widely used because they are simple to compute. Nevertheless, further research is needed to provide guidelines on the choice of a weighting scheme for the weighted kappa coefficient.

The continuation of this chapter is a review of methods to compare several kappa-like coefficients or more generally to study the effect of a set of predictors (characteristics of the subjects and/or the raters) on the agreement level in order to provide researchers practical means to improve the agreement level between raters (see e.g. Williamson et al. (2000); Vanbelle and Albert (2008); Vanabelle et al. 2012)).

Appendix: Variance of the Kappa Coefficients

The sample variance of Cohen's kappa coefficient using the delta method is given by

$$\mathrm{var}(\hat{\kappa}) = \frac{p_o(1-p_o)}{N(1-p_e)^2} + \frac{2(p_o-1)(C_1-2p_op_e)}{N(1-p_e)^3} + \frac{(p_o-1)^2(C_2-4p_e^2)}{N(1-p_e)^4} \tag{1.5}$$

where

$$C_1 = \sum_{j=1}^{K} p_{jj}(p_{j.}+p_{.j}) \quad \text{and} \quad C_2 = \sum_{j=1}^{K}\sum_{k=1}^{K} p_{jk}(p_{.j}+p_{k.})^2.$$

With the additional assumption of no rater bias, the sample variance simplifies to

$$\mathrm{var}(\hat{\kappa}_I) = \frac{1}{N(1-C_3)^2}\left\{\begin{array}{l}\displaystyle\sum_{j=1}^{K} p_{jj}[1-4\bar{p}_j(1-\hat{\kappa}_I)] \\ \displaystyle +(1-\hat{\kappa}_I)^2\sum_{j=1}^{K}\sum_{k=1}^{K} p_{jk}(\bar{p}_j+\bar{p}_k)^2-[\hat{\kappa}_I-C_3(1-\hat{\kappa}_I)]^2\end{array}\right\} \tag{1.6}$$

where $\bar{p}_j = (p_{j.}+p_{.j})/2$ and $C_3 = \sum_{j=1}^{K} \bar{p}_j$ $(j,k = 1,\cdots,K)$.

The two sided $(1-\alpha)$ confidence interval for κ is then determined by $\hat{\kappa} \pm Q_z(1-\alpha/2)\sqrt{\text{var}(\hat{\kappa})}$, where $Q_z(1-\alpha/2)$ is the $\alpha/2$ upper percentile of the standard Normal distribution.

The sample variance of the weighted kappa coefficient obtained by the delta method is

$$\text{var}(\hat{\kappa}_w) = \frac{1}{N(1-p_{ew})^4} \left\{ \begin{array}{l} \sum_{j=1}^{K}\sum_{k=1}^{K} p_{jk}\left[w_{jk}(1-p_{ew}) - (\overline{w}_{j.} + \overline{w}_{.k})(1-p_{ow})\right]^2 \\ -(p_{ow}p_{ew} - 2p_{ew} + p_{ow})^2 \end{array} \right\} \quad (1.7)$$

where $\overline{w}_{.j} = \sum_{m=1}^{K} w_{mj}p_{m.}$ and $\overline{w}_{k.} = \sum_{s=1}^{K} w_{ks}p_{.s}$.

Bibliography

Byrt T, Bishop J, Carlin J (1993) Bias, prevalence and kappa. J Clin Epidemiol 46:423–429

Cicchetti DV, Allison T (1971) A new procedure for assessing reliability of scoring EEG sleep recordings. Am J EEG Technol 11:101–109

Cohen J (1960) A coefficient of agreement for nominal scales. Educ Psychol Meas 20:37–46

Cohen J (1968) Weighted kappa: nominal scale agreement with provision for scaled disagreement or partial credit. Psychol Bull 70:213–220

Fermanian J (1984) Mesure de l'accord entre deux juges. Cas qualitatif. Rev d'Epidemiol St Publ 32:140–147

Fleiss JL, Cohen J (1973) The equivalence of weighted kappa and the intraclass correlation coefficient as measure of reliability. Educ Psychol Meas 33:613–619

Gilmour E, Ellerbrock TV, Koulos JP, Chiasson MA, Williamson J, Kuhn L, Wright TC Jr (1997) Measuring cervical ectopy: direct visual assessment versus computerized planimetry. Am J Obstet Gynecol 176:108–111

Kraemer HC (1979) Ramifications of a population model for κ as a coefficient of reliability. Psychometrika 44:461–472

Landis JR, Koch GG (1977) The measurement of observer agreement for categorical data. Biometrics 33:159–174

Schuster C (2004) A note on the interpretation of weighted kappa and its relation to other rater agreement statistics for metric scales. Educ Psychol Meas 64:243–253

Scott WA (1955) Reliability of content analysis: the case of nominal scale coding. Public Opin Q 19:321–325

Thompson WD, Walter SD (1988) A reappraisal of the kappa coefficient. J Clin Epidemiol 41:949–958

Vanbelle S, Albert A (2008) A bootstrap method for comparing correlated kappa coefficients. J Stat Comput Simul 78:1009–1015

Vanbelle S, Albert A (2009) A note on the linearly weighted kappa coefficient for ordinal scales. Stat Methodol 6:157–163

Vanbelle S, Mutsvari T, Declerck D, Lesaffre E (2012) Hierarchical modeling of agreement. Stat Med 31(28):3667–3680

Warrens M (2012) Some paradoxical results for the quadratically weighted kappa. Psychometrika 77:315–323

Williamson JM, Lipsitz SR, Manatunga AK (2000) Modeling kappa for measuring dependent categorical agreement data. Biostatistics 1:191–202

Chapter 2
Clinical Agreement in Quantitative Measurements

Limits of Disagreement and the Intraclass Correlation

Abhaya Indrayan

Abstract In clinical research, comparison of one measurement technique with another is often needed to see whether they agree sufficiently for the new to replace the old. Such investigations are often analysed inappropriately, notably by using correlation coefficients, which could be misleading. This chapter describes alternatives based on graphical techniques that quantify disagreement as well as the concept of intraclass correlation.

Introduction

Assessment of agreement between two or more measurements has become important for the following reasons. Medical science is growing at a rapid rate. New instruments are being invented and new methods are being discovered that measure anatomic and physiologic parameters with better accuracy and precision, and at lower cost. Emphasis is on simple, non-invasive, safer methods that require smaller sampling volumes and can help in continuous monitoring of patients when required. Acceptance of any new method depends on a convincing demonstration that it is nearly as good as, if not better than, the established method. The problem in this case is not equality of averages but of equality of all individual values.

The term agreement is used in several different contexts. The following discussion is restricted to a setup where a pair of observations (x,y) is obtained by measuring the same characteristic on the same subject by two different methods, by two different observers, by two different laboratories, at two anatomic sites, etc. There can also be more than two. The measurement could be qualitative or quantitative. Quantitative agreement is between exact values, such as intra-ocular

A. Indrayan (✉)
Department of Biostatistics and Medical Informatics, University College of Medical Sciences, Delhi, India
e-mail: a.indrayan@gmail.com

S.A.R. Doi and G.M. Williams (eds.), *Methods of Clinical Epidemiology*,
Springer Series on Epidemiology and Public Health,
DOI 10.1007/978-3-642-37131-8_2,

pressure in two eyes, and quantitative agreement is between attributes such as the presence or absence of a minor lesion in radiographs read by two radiologists. The method of assessing agreement in these two cases is different. This chapter is on agreement in quantitative measurements. Agreement in qualitative measurements is discussed in the previous chapter.

Assessment of Quantitative Agreement

Irrespective of what is being measured, it is highly unlikely that the new method would give exactly the same reading in every case as the old method, even if they are equivalent. Some differences would necessarily arise – if nothing else, at least as many as would occur when the same method is used two times on the same subject under identical conditions. How do you decide that the new method is interchangeable with the old? The problem is described as one of quantitative agreement. This is different from evaluating which method is better. The assessment of better is done with reference to a gold standard. Assessment of agreement does not require any such standard.

Quantitative Measurements

The problem of agreement in quantitative measurement can arise in at least five different types of situations. (1) Comparison of self-reported values with instrument-measured values, for example, urine frequency and bladder capacity using a patient questionnaire and a frequency–volume chart. (2) Comparison of measurements at two or more different sites, for example, paracetamol concentration in saliva with that in serum. (3) Comparison of methods, for example, bolus and infusion methods of estimating hepatic blood flow in patients with liver disease. (4) Comparison of two observers, for example, duration of electroconvulsive fits reported by two or more psychiatrists on the same group of patients, or comparison of two or more laboratories when, for example, aliquots of the same sample are sent to two different laboratories for analysis. (5) Intraobserver consistency, for example, measurement of the anterior chamber depth of an eye segment two or more times by the same observer using the same method to evaluate the reliability of the method.

In the first four cases, the objective is to find whether a simple, safe, less expensive procedure can replace an existing procedure. In the last case, the reliability of the method is being evaluated.

Statistical Formulation of the Problem

The statistical problem in all these cases is to check whether or not a $y = x$ type of relationship exists in individual subjects. This looks like a regression setup $y = a + bx$ with $a = 0$ and $b = 1$, but that is not really. The difference is that, in regression, the relationship is between x and the average of y. In an agreement setup, the concern is with individual values and not with averages. Nor should agreement be confused with high correlation. Correlation is nearly 1 if there is a systematic bias and nearly same difference occurs in every subject. Example 1 illustrates the distinction between $y = x$ regression and agreement.

Example : Very Different Values but Regression Is y = x

The following Hb values are reported by two laboratories for the same blood samples:

Lab I (x)	11.3	12.0	13.9	12.8	11.3	12.0	13.9	12.8
Lab II (y)	11.5	12.4	14.2	13.2	11.1	11.6	13.6	12.4

$$\bar{x} = 12.5, \quad \bar{y} = 12.5, \quad r = 0.945$$

$$\hat{y} = x, \quad \text{that is,} \ b = 1 \quad \text{and} \quad a = 0$$

The two laboratories have the same mean for these eight samples and a very high correlation (0.945). The intercept is 0 and slope is 1.00. Yet there is no agreement in any of the subjects. The difference or error ranges from 0.2 to 0.4 g/dL. This is substantial in the context of the present-day technology for measuring Hb levels. Thus, equality of means, a high degree of correlation and regression $y = x$ are not enough to conclude agreement. Special methods are required.

The first four values of x in this example are the same as the last four values. The first four values of y are higher and the last four values are lower by the same margin. Thus, for each x, $\bar{y} = x$ giving rise to the regression $\hat{y} = x$. In this particular case, the correlation coefficient is also nearly 1.

Quantitative agreement in individual values can be measured either by limits of disagreement or by intraclass correlation. The details are as follows.

Limits of Disagreement Approach

This method is used for a pair of measurements and based on the differences $d = (x - y)$ in the values obtained by the two methods or observers under comparison. If the methods are in agreement, this difference should be zero for every subject. If these differences are randomly distributed around zero and none of the differences is large, the agreement is considered good. A graphical approach is to plot d versus $(x + y)/2$. A flat line around zero is indicative of good agreement. Depending on which is labelled x and which is y, an upward trend indicates that x is generally more than y, and a downward trend that y is more than x.

A common sense approach is to consider agreement as reasonably good if, say, 95 % of these differences fall within the prespecified clinically tolerable range and the other 5 % are also not too far from that. Statistically, when the two methods or two observers are measuring the same variable, then the difference d is mostly the measurement error. Such errors are known to follow a Gaussian distribution. Thus the distribution of d in most cases would be Gaussian. Then the limits $\bar{d} \pm 1.96s_{\mathrm{d}}$ are likely to cover differences in nearly 95 % of subjects where $\bar{d}$ is the average and s_{d} is the standard deviation (SD) of the differences. The literature describes them as the limits of agreement. They are actually limits of disagreement.

$$\text{Limits of disagreement: } \bar{d} - 1.96s_{\mathrm{d}} \quad \text{to} \quad \bar{d} + 1.96s_{\mathrm{d}} \tag{2.1}$$

If these limits are within clinical tolerance in the sense that a difference of that magnitude does not alter the management of the subjects, then one method can be replaced by the other. The mean difference $\bar{d}$ is the bias between the two sets of measurements and s_{d} measures the magnitude of random error. For further details, see Bland and Altman (1986).

The limitations of the product–moment correlation coefficient are well known. Consider the following example. Suppose a method consistently gives a level 0.5 mg/dL higher than another method. The correlation coefficient between these two methods would be a perfect 1.0. Correlation fails to detect systematic bias. This also highlights the limitations of the limits of disagreement approach. The difference between measurements by two methods is always +0.5 mg/dL, thus the SD of the difference is zero. The limits of disagreement in this case are (+0.5,+0.5). This is in fact just one value and not limits. A naive argument could be that these limits are within clinical tolerance and thus the agreement is good. To detect this kind of fallacy, plot the differences against the mean of paired values. This plot can immediately reveal this kind of systematic bias.

Table 2.1 Results on agreement between AVRG[a] and Korotkoff BP readings in 100 volunteers

	AVRG[a]		Korotkoff
Mean systolic BP (mmHg)	115.1		115.5
SD (mmHg)	13.4		13.2
Mean difference (mmHg)		−0.4	
P-value for paired t		>0.50	
Correlation coefficient (r)		0.87	
SD of difference, s_d (mmHg)		6.7	
Limits of disagreement (mmHg)		(−13.5, 12.7)	
Intraclass correlation coefficient (r_I) (formula given in next section)		0.87	

[a]Average of readings at the appearance and disappearance of the plethysmographic waveform of a pulse oximeter

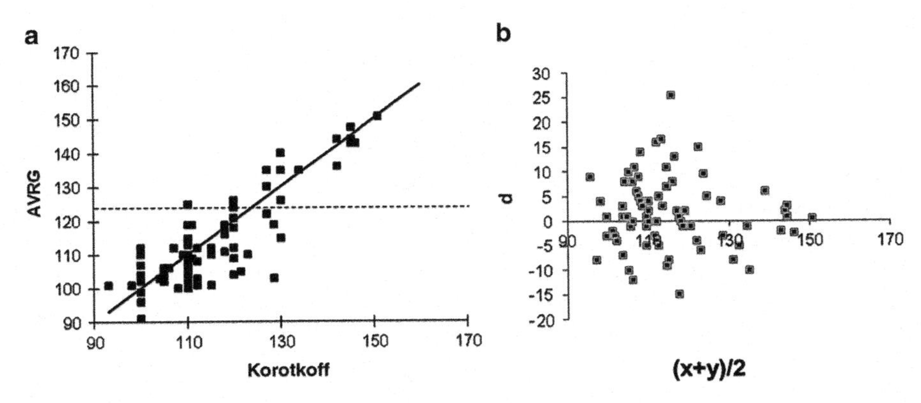

Fig. 2.1 (**a**) Scatter of the pulse oximeter based (y) and Korotkoff based (x) readings of systolic blood pressure. For pulse oximeter based readings the average of readings at the disappearance and reappearance of the waveform respectively were used (labelled AVRG in the *left panel*). (**b**) Plot of d versus $(x + y)/2$ (d = difference between y and x which are defined as above)

Example : Limits of Disagreement Between Pulse Oximetry and Korotkoff Readings

Consider the study by Chawla et al. (1992) on systolic blood pressure (BP) readings derived from the plethysmographic waveform of a pulse oximeter. This method could be useful in a pulseless disease such as Takayasu syndrome. The readings were obtained (a) at the disappearance of the waveform on the pulse oximeter on gradual inflation of the cuff and (b) at the reappearance on gradual deflation. In addition, BP was measured in a conventional manner by monitoring the Korotkoff sounds. The study was done on 100 healthy volunteers. The readings at disappearance of the waveform were generally higher and at reappearance generally lower. Thus, the average (AVRG) of the two is considered a suitable value for investigating the agreement with the Korotkoff readings. The results are shown in Table 2.1. The scatter, the line of equality and the plot of d versus $(x + y)/2$ are shown in Fig. 2.1. Figure 2.1b shows that the differences were large for smaller values.

Despite the means being nearly equal and r very high, the limits of disagreement (Table 2.1) show that a difference of nearly 13 mmHg can arise between the two readings on either side (average of pulse oximetry readings can give either less or more than the Korotkoff readings). These limits are further subject to sampling fluctuation, and the actual difference in individual cases can be higher. Now it is for the clinician to decide whether a difference of such magnitude is tolerable. If it is, then the agreement can be considered good and pulse oximetry readings can be used as a substitute for Korotkoff readings, otherwise they should not be used. Thus, the final decision is clinical rather than statistical when this procedure is used.

Intraclass Correlation as a Measure of Agreement

Intraclass correlation is the strength of a linear relationship between subjects belonging to the same class or the same subgroup or the same family. In the agreement setup, the two measurements obtained on the same subject by two observers or two methods is a subgroup. If they agree, the intraclass correlation will be high. This method of assessing an agreement was advocated by Lee et al. (1989).

In the usual correlation setup, the values of two different variables are obtained on a series of subjects. For example, you can have the weight and height of 20 girls aged 5–7 years. You can also have the weight of the father and mother of 30 low birthweight newborns. Both are weights and the product–moment correlation coefficient is a perfectly valid measure of the strength of the relationship in this case. Now consider the weight of 15 persons obtained on two machines. Any person, say number 7, may be measured by machine 2 first and then by machine 1. Others may be measured by machine 1 then by machine 2. The order does not matter in this setup as the interest is in finding whether the values are in agreement or not.

Statistically, intraclass correlation is that part of the total variance that is accounted for by the differences in the paired measurements obtained by two methods. That is,

$$\text{Intraclass correlation:} \quad \rho_{\text{I}} = \frac{\sigma_{\text{M}}^2}{\sigma_{\text{M}}^2 + \sigma_{\text{e}}^2} \tag{2.2}$$

where σ_{M}^2 is the variance between methods if methods are to be compared for agreement and σ_{e}^2 is the error variance. This formulation does not restrict us to only two methods. These could be three or more. In the weight example, you can compare agreement among five machines by taking the weight of each of the 15 persons on these five machines.

The estimate of ρ_{I} is easily obtained by setting up the usual analysis of variance (ANOVA) table. If there are M methods under comparison, the ANOVA table would look like Table 2.2. The number of subjects is n in this table and other notations are self-explanatory. E(MS) is the expected value of the corresponding mean square.

Table 2.2 Structure of ANOVA table in agreement setup

Source	df	Mean square (MS)	E(MS)
Methods (A)	$M-1$	MSA	$\sigma_e^2 + n\sigma_M^2$
Subjects (B)	$n-1$	MSB	$\sigma_e^2 + M\sigma_S^2$
Error	$(M-1)(n-1)$	MSE	σ_e^2

A little algebra yields the estimate of the intraclass correlation r_I:

$$r_I = \frac{\text{MSA} - \text{MSE}}{\text{MSA} + (n-1)\text{MSE}} \tag{2.3}$$

This can be easily calculated once you have the ANOVA table. Statistical software will give you the value of the intraclass correlation directly.

In terms of the available values, the computation of the intraclass correlation coefficient (ICC) is slightly different from that of the product–moment correlation coefficient. In the agreement setup, the interest is in the correlation between two measurements obtained on the same subject and is obtained as follows.

ICC (a pair of readings):

$$r_I = \frac{2\Sigma_i(x_{i1} - \bar{x})(x_{i2} - \bar{x})}{\Sigma_i(x_{i1} - \bar{x})^2 + \Sigma_i(x_{i2} - \bar{x})^2} \tag{2.4}$$

where x_{i1} is the measurement on the ith subject ($i = 1,2,\ldots,n$) when obtained by the first method or the first observer, x_{i2} is the measurement on the same subject by the second method or the second observer, and $\bar{x}$ is the overall mean of all $2n$ observations. Note the difference in the denominator compared with the formula for the product–moment correlation.

This was calculated for the systolic BP data described in Example 2 and was found to be $r_I = 0.87$. A correlation >0.75 is generally considered enough to conclude good agreement. Thus, in this case, the conclusion on the basis of the intraclass correlation is that the average of readings at disappearance and appearance of the waveform in pulse oximetry in each person agrees fairly well with the Korotkoff readings for that person. This may not look consistent with the limits of disagreement that showed a difference up to 13 mmHg between the two methods. The two approaches of assessing agreement can sometimes lead to different conclusions.

Equation (2.4) is used for comparing two methods or two raters. This correlation can be used for several measurements. For example, you may have the wave amplitude of electrical waves at $M = 6$ different sites in the brain of each of $n = 40$ persons. For multiple raters or multiple methods, ICC (several readings):

$$r_I = \frac{\Sigma_i\Sigma_{j\neq k}(x_{ij} - \bar{x})(x_{ik} - \bar{x})}{(M-1)\Sigma_i\Sigma_j(x_{ij} - \bar{x})^2}, \quad i = 1,2,\ldots,n, \quad j,k = 1,2,\ldots,M \tag{2.5}$$

where n is the number of subjects and M is the number of observers or the number of methods to be compared. The mean $\bar{x}$ is calculated on the basis of all Mn observations.

Table 2.3 Cutoffs for grading the strength of agreement

Intraclass correlation	Strength of agreement
<0.25	Poor
0.25–0.50	Fair
0.50–0.75	Moderate
0.75–0.90	Good
>0.90	Excellent

For grading of the strength of agreement, the cutoffs shown in Table 2.3 can be used.

An Alternative Simple Approach to Agreement Assessment

Neither of the two methods described in the preceding sections is perfect. Let us first look at their relative merits and demerits and then propose an alternative method, which may also not be perfect but is relatively simple.

Relative Merits of the Two Methods

Indrayan and Chawla (1994) studied the merits and demerits of the two approaches in detail. The following are their conclusions on the comparative features of the two methods:

1. The ICC does not depend on the subjective assessment of any clinician. Thus, it is better to base the conclusion on this correlation when the clinicians disagree on the tolerable magnitude of differences between two methods (or two observers). And clinicians seldom agree on such issues.
2. The 0.75 threshold to label an intraclass correlation high or low is arbitrary, although generally acceptable. Thus, there is also a subjective element in this approach.
3. Intraclass correlation is unit free, easy to communicate, and interpretable on a scale of zero (no agreement) to one (perfect agreement). This facility is not available in the limits of disagreement approach.
4. A distinct advantage of the limits of disagreement approach is its ability to delineate the magnitude of individual differences. It also provides separate estimates of bias $(\bar{d})$ and random error (s_d). This bias measures the constant differences between the two measurements and random error is the variation around this bias. Also, this approach is simple and does not need much calculation.
5. The limits of disagreement can be evaluated only when the comparison is between two measurements. The intraclass correlation, on the other hand, is fairly general and can be used for comparing more than two methods or more than two observers (Eq. 2.5).

6. Intraclass correlation can also be used for comparing one group of raters with another group. Suppose you have four male assessors and three female assessors. Each subject is measured by all seven assessors. You can compare intraclass correlation obtained for male assessors with that obtained for female assessors. You can have one set of subjects for assessment by males and another set of subjects for assessment by females.

A review of the literature suggests that researchers prefer the limits of disagreement approach to the ICC approach for comparing two methods. A cautious approach is to use both and come to a firm conclusion if both give the same result. If they are in conflict, defer a decision and carry out further studies.

The following comments might provide better appreciation of the procedure to assess quantitative agreement:

1. As mentioned earlier, the limits of disagreement $\bar{d} \pm 1.96 s_{\mathrm{d}}$ themselves are subject to sampling fluctuation. A second sample of subjects may give different limits. Methods are available to find an upper bound to these limits. For details, see Bland and Altman (1986). They call them limits of agreement, but perhaps they should be called limits of disagreement.
2. The ICC too is subject to sampling fluctuation. For assessing agreement, the relevant quantity is the lower bound of r_{I}. This can be obtained by the method described by Indrayan and Chawla (1994). Their method for computing the ICC is based on ANOVA, but that gives the same result as obtained by Eq. (2.4).
3. Although not specifically mentioned, the intraclass correlation approach assumes that the methods or observers under comparison are randomly chosen from a population of methods or observers. This is not true when comparing methods because they cannot be considered randomly chosen. Thus, the intraclass correlation approach lacks justification in this case. However, when comparing observers or laboratories, the assumption of a random selection may have some validity. If observers or laboratories agree, a generalized conclusion about consistency or reliability across them can be drawn.
4. Intraclass correlation is also used to measure the reliability of a method of measurement as discussed briefly by Indrayan (2012).
5. Both these approaches are applicable when both the methods could be in error. As mentioned earlier, these methods are not appropriate for comparing with a gold standard that gives a fixed target value for each subject. For agreement with a gold standard, see Lin et al. (2002).

An Alternative Simple Approach

The limits of disagreement approach just described is based on the average difference and has the limitations applicable to all averages. For example, this approach does not work if the bias or error is proportional. Fasting blood glucose levels vary from 60 to 300 mg/dL or more. Five percent of 60 is 3 and of 300 is 15. The limits of

Table 2.4 Data on fasting blood sugar levels in 10 blood samples

Method 1 (x)	86	172	75	244	97	218	132	168	118	130
Method 2 (y)	90	180	73	256	97	228	138	172	116	132
$d = x - y$	−4	−8	+2	−12	0	−10	−6	−4	+2	−2
5 % of x	4.30	8.60	3.75	12.20	4.85	10.90	6.60	8.40	5.90	6.50

disagreement approach considers them to be different and ignores that both are 5 % and proportionately the same. Also, if one difference is 10 and the other is 2, and they are not necessarily proportional, the limits of disagreement consider only the average. Individual differences tend to be overlooked. A few unusually large differences distort the average and are not properly accounted except by disproportional inflation of the SD.

To account for small and big individual differences as well as proportional bias, it may be prudent to set up a clinical limit that can be tolerated for individual differences without affecting the management of the condition. Such limits are required anyway for the limits of disagreement approach, albeit for the average. These clinical limits of indifference can be absolute or in terms of a percentage. If not more than a prespecified percentage (say 5 %) of individual differences are beyond these limits in a large sample, you can safely assume adequate agreement. This does not require any calculation of the mean and SD. You may like to add a condition such as none of the differences should be more than two times the limit of indifference. Any big difference, howsoever isolated, raises alarm. A plot of y versus x can track that the differences are systematic or random.

Example : Agreement Between Two Methods of Measuring Fasting Blood Glucose Levels

Consider the data in Table 2.4. Suppose method 1 is the current standard although this can also be in error. Method 2 is extremely cheap and gives instant results. Suppose also that clinicians are willing to accept 5 % error in view of the distinct advantages of method 2. Note that this indifference is a percentage and not an absolute value.

In these data, the y versus x plot is on a fairly straight line (Fig. 2.2a) but the plot of d versus $(x + y)/2$ (Fig. 2.2b) shows an aberration with a large number of points on the negative side and following an increasing trend. This shows lack of agreement according to the limits of disagreement approach. This really is not the case as explained next.

None of the differences exceed the clinical limit of indifference of 5 % in this sample. Thus, method 2 can be considered in agreement with method 1 although a larger sample is required to be confident. However, most differences are negative, indicating that method 2 generally provides lower values. The average difference is 4.2 mg/dL in absolute terms and nearly 3 % of y in relative terms. This suggests the correction factor for bias. If you decide to subtract 3 % of the level obtained by method 2, you can reach very close to the value obtained by method 1 in most cases. Do this as an exercise and verify it for yourself.

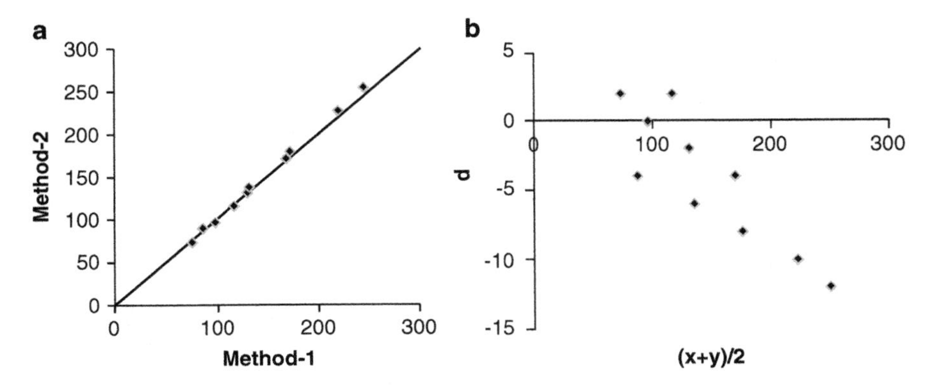

Fig. 2.2 (**a**) y versus x plot for data in Example 3, and (**b**) d versus $(x + y)/2$ plot for the same data. The variables x and y are the results of glucose measurements on the same sample by two different methods (mg/dl) and the difference between the results of two methods is given by d

Now forget about 5 % tolerance, and note that some differences are small and some are quite large in Example 3. The value of $s_d = 4.85$ in this case. Thus, the limits of disagreement are $-4.2 \pm 2 \times 4.85$, or -13.9 to +5.5. These limits may look too wide and beyond clinical tolerance, particularly on the negative side. These limits do not allow a larger error for larger values that proportionate considerations would allow. Also, these are based on an average and do not adequately consider individual differences. If 1 out of 20 values shows a big difference, this can distort the mean and inflate the SD, and provide unrealistic limits of disagreement. The alternative approach suggested above can be geared to allow not more than 5 % individual differences beyond the tolerance limit and you can impose an additional condition that none should exceed, say, by 10 % of the base value. Since it is based on individual differences and not on an average, this alternative approach may be more appealing.

Bibliography

Bland JM, Altman DG (1986) Statistical methods for assessing agreement between two methods of clinical measurement. Lancet 1:307–310

Chawla R, Kumarvel V, Girdhar KK, Sethi AK, Indrayan A, Bhattacharya A (1992) Can pulse oximetry be used to measure systolic blood pressure? Anesth Analg 74:196–200

Indrayan A (2012) Medical biostatistics, 3rd edn. Chapman & Hall/CRC Press, Boca Raton

Indrayan A, Chawla R (1994) Clinical agreement in quantitative measurements. Natl Med J India 7:229–234

Lee J, Koh D, Ong CN (1989) Statistical evaluation of agreement between two methods for measuring a quantitative variable. Comput Biol Med 19:61–70

Lin L, Hedayat AS, Sinha B, Yang M (2002) Statistical methods in assessing agreement: models, issues and tools. J Am Stat Assoc 7:257–270

Chapter 3
Disagreement Plots and the Intraclass Correlation in Agreement Studies

Suhail A.R. Doi

Abstract Although disagreement and the intraclass correlation have been covered previously, several variants of both have been proposed. This chapter introduces readers to several variants of the disagreement plot and the classification of the intraclass correlation coefficient and the concept of repeatability in agreement studies.

Variations of the Limits of Disagreement Plot

Several variations have been proposed for the limits of disagreement plot (also called a Bland–Altman plot). In terms of the x-axis, instead of the average we could use a geometric mean or even the values on one of the two methods, if this is a reference or gold standard method (see Krouwer 2008). In terms of the y-axis, the difference (d) can be expressed as percentages of the values on the average of the measurements (i.e. proportional to the magnitude of measurements). This helps when there is an increase in variability of the differences as the magnitude of the measurement increases. Ratios of the measurements can be plotted instead of d (avoiding the need for log transformation). This option is also utilized when there is an increase in variability of the differences as the magnitude of the measurement increases. All these can be done with the help of routine software such as MedCalc, which gives a warning when either the ratio or percentage includes zero values.

S.A.R. Doi (✉)
Clinical Epidemiology Unit, School of Population Health, University of Queensland, Brisbane, Australia

Princess Alexandra Hospital, Brisbane, Australia
e-mail: sardoi@gmx.net

S.A.R. Doi and G.M. Williams (eds.), *Methods of Clinical Epidemiology*, Springer Series on Epidemiology and Public Health,
DOI 10.1007/978-3-642-37131-8_3,

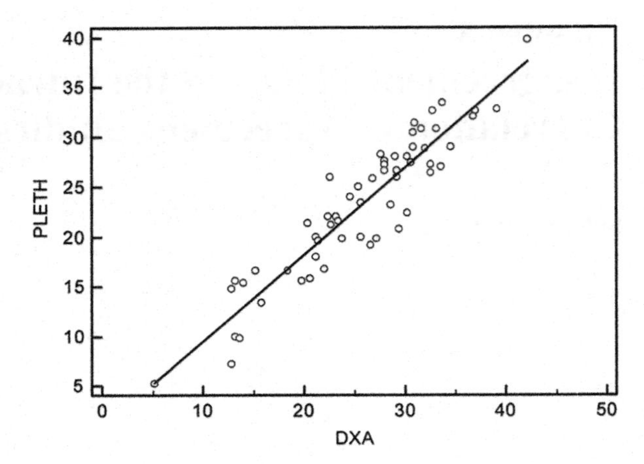

Fig. 3.1 Regression of DXA based %body fat versus PLETH based %body fat demonstrating a good linear relationship. However this does not necessarily mean there is good agreement

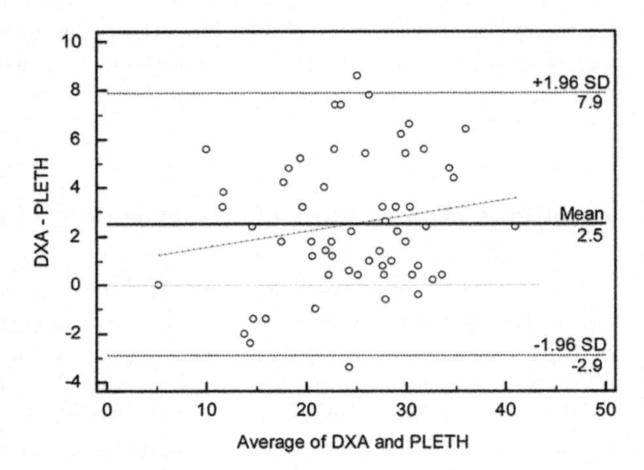

Fig. 3.2 The classic layout of the Bland–Altman plot comparing %body fat measurements by DXA and PLETH in terms of limits of disagreement. The limits suggest that measuring % body fat by PLETH could result in an absolute difference of −2.9 % to +7.9 % compared with DXA

Example

Sardinha (1998) designed a study to compare air displacement plethysmography with dual-energy X-ray absorptiometry (DXA) for estimation of percent body fat (%BF). The differences between DXA-determined and plethysmography-determined %BF were compared (Figs. 3.1, 3.2, 3.3, 3.4, and 3.5).

The Special Case of Limits of Disagreement for two Successive Measurements: Repeatability Coefficient

The repeatability coefficient tells us the maximum difference likely to occur between two successive measurements when we want to know how good a measurement instrument is.

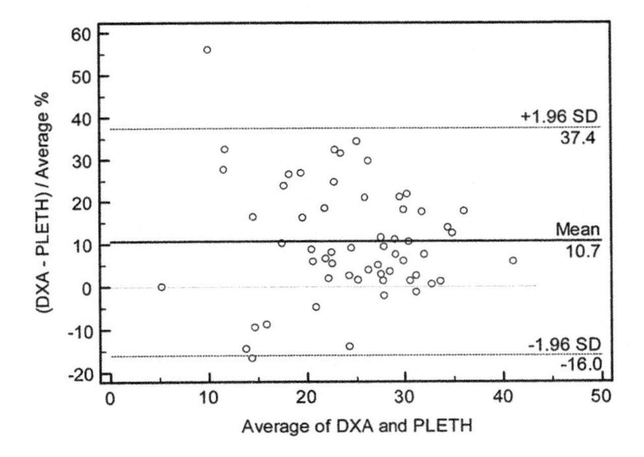

Fig. 3.3 The % average layout of the Bland–Altman plot comparing $body fat via DXA with %body fat via PLETH in terms of limits of disagreement. The limits suggest that measuring %body fat by PLETH could result in a relative difference of −16 % to +37.4 % compared with DXA. This option is useful when there is an increase in variability of the differences as the magnitude of the measurement increases

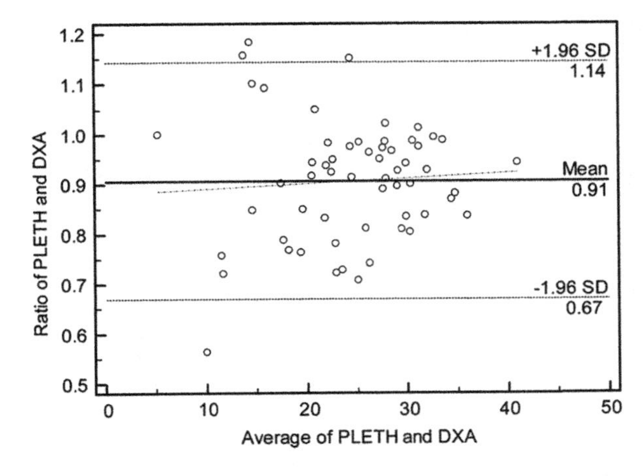

Fig. 3.4 The ratio layout of the Bland–Altman plot comparing %body fat by DXA with %body fat by PLETH in terms of limits of disagreement. The limits suggest that measuring %body fat by PLETH could result in a value 0.67–1.14 times that of DXA. Again, this option is useful when there is an increase in variability of the differences as the magnitude of the measurement increases. However, there must not be a zero value for either one of the two techniques

Assessment of reproducibility requires starts with measurements per subject as given in the example (Table 3.1) of glucometer readings in 15 subjects. If the variance of the differences between the 15 pairs of repeated measurements = 0.1303, the repeatability is then given by:

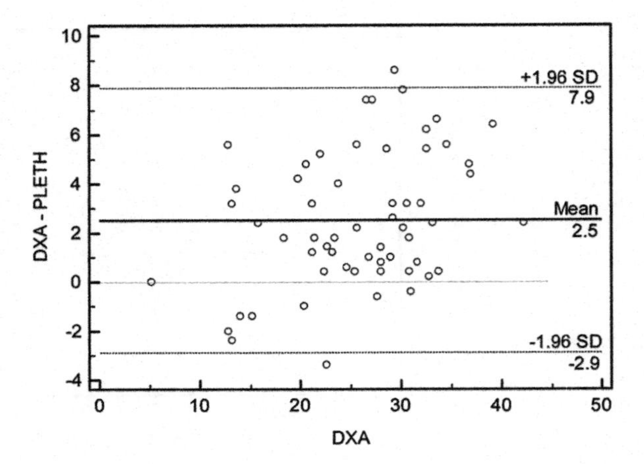

Fig. 3.5 Classic layout of the Bland–Altman plot but using only DXA measurements on the horizontal axis rather than the average of DXA and PLETH measurements. Here again, %body fat by PLETH can be an absolute 2.5 % less to 7.9 % more than DXA. This option is utilized if one of the measurements is a gold standard

Table 3.1 Differences between glucometer readings on a sample taken from each subject

Subject	Glucometer type 1	Glucometer type 2	Difference between readings
1	4.9	5.4	−0.5
2	4.0	4.0	0
3	5.2	5.2	0
4	4.3	4.4	−0.1
5	4.8	5.3	−0.5
6	5.6	5.9	−0.3
7	4.1	4.1	0
8	4.4	5.3	−0.9
9	6.5	7.6	−1.1
10	4.3	5.0	−0.7
11	4.2	4.5	−0.3
12	6.6	7.5	−0.9
13	2.7	2.8	−0.1
14	4.8	5.4	−0.6
15	1.8	2.1	−0.3

$$\text{repeatability} = \sqrt{0.1303} \times 1.96 = 0.71$$

This impiles that for the glucometers, for 95 % of all pairs of measurements on the same subject, the difference between two replicates may be as much as 0.71 mmol/L just by random measurement error alone. Bland and Altman called this value of 0.71 mmol/L the repeatability of measurement. Bland and Altman have proposed this as equivalent to the limits of disagreement for repeated measures.

We can get a close approximation to the variance of the difference from a simpler expression:

$$\frac{\sum d^2}{2N}$$

where d is the difference between successive measurements and N is the number of pairs of repeated measures. Thus repeatability is the given by:

$$1.96 \times \sqrt{\frac{\sum d^2}{2N}}$$

If we use this approximation, it slightly overestimates the repeatability.

Given that the variance of the difference equals the sum of the within subject variances of each measurement, repeatability can also be defined in terms of the within subject variance (σ^2_{within}) as follows:

$$1.96 \times \sqrt{2\sigma^2_{\text{within}}} = 1.96 \times \sqrt{2} \times \sigma^2_{\text{within}} = 2.77 \times \sigma^2_{\text{within}}$$

2.77 is called the repeatability coefficient by Bland and Altman.

Further Details on the Intraclass Correlation Coefficient

Use in Rater Evaluation

Results for each sample or subject tested more than once may be considered a cluster or a class of measurements by different methods or raters. The intraclass correlation coefficient (ICC) is thus a measure of agreement within these clusters or classes. For example, if the clusters (or classes) are made up of different physicians' ratings for individual papers (i.e. they give a quality score to each research paper) then this can be depicted as shown in Figs. 3.6 and 3.7.

Take the extreme case where each paper receives the same score from both raters; i.e. no variance within the raters (Fig. 3.6). So, ICC = variance between class/(variance between class + variance within class) = $\sigma_B{}^2/[\sigma_B{}^2 + \sigma^2{}_W]$ = $\sigma^2{}_B/[\sigma_B{}^2 + 0] = 1$ = perfect agreement within the cluster (class). Thus, papers (class/cluster) are more heterogeneous than ratings within clusters (or class). Alternatively, ratings are more homogeneous (good raters) than the class/cluster.

A different case arises when each paper has raters who give very different scores (Fig. 3.7); that is, most of the variance is within the raters (difference between papers (class) are now less than within their ratings). ICC is close to 0 implying bad raters.

When agreement is assessed for repeated measurements or ratings, the data set for analysis of agreement between methods or raters is usually laid out such that the rows represent each class/cluster and the columns contain the different measurements that make up each class/cluster. In general, the ICC approaches 1.0 as the between-class effect (the row effect) is very large relative to the within-class effect (the column effect), regardless of what the rows and columns represent. In this way, the ICC is a measure of homogeneity within class; it approaches 1.0 when any given row tends to have the same values for all columns.

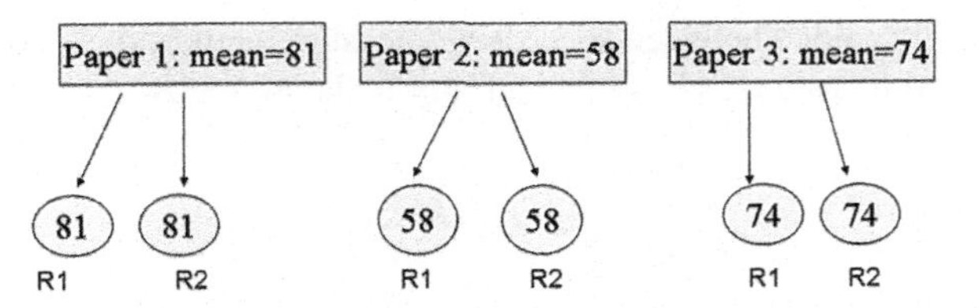

Fig. 3.6 Different physicians' ratings for individual papers. The ovals are the raters' and the rectangles the papers (which form the class or cluster). Each paper has a different average score

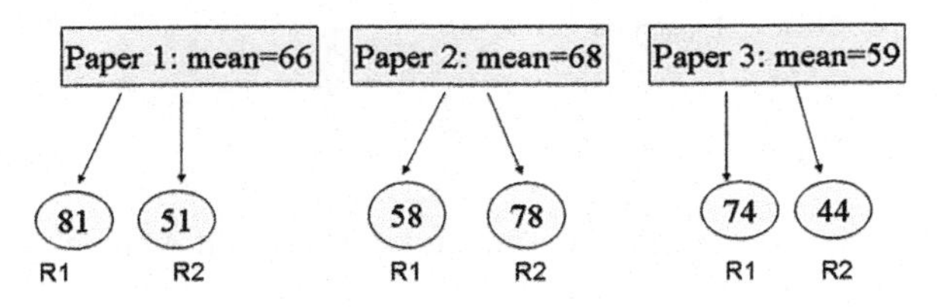

Fig. 3.7 Different physicians' ratings for individual papers. The ovals are the raters' and the rectangles the papers (which form the class or cluster). Each paper has a similar average score

To take an example, let columns be different glucometers and let rows be subject samples, and let the attribute measured be blood glucose. If glucometers vary much more than the glucose levels across samples, then ICC will be low. When the ICC is low, most of the variation is within the class/cluster (columns – glucometers). In this case the across class/cluster differences (rows – samples) are less than the within class/cluster differences (columns – glucometers). In terms of multivariable analysis of glucometer results, there is less difference between the results from ordinary regression and multilevel models when the ICC is low. In such cases, it may be less important to use a statistical model that allows variables for within-cluster characteristics, for example, different types of glucometers. The ICC is 0.5 when within-group variance equals between-group variance, indicative of the glucometer having no effect. Although less common, note that the ICC can become negative when the within-group variance exceeds the between-group variance.

Classification of ICCs

Based on the rating design, Shrout and Fleiss (1979) defined three types of ICCs (types 1–3):

1. ICC(1,*k*): each target (e.g. research paper) is rated by different raters. Absolute agreement (one-way random effects).
2. ICC(2,*k*): the same but exchangeable raters rate each target. Absolute agreement, because systematic differences are relevant (two-way random effects).

3. ICC(3,*k*): the same but non-exchangeable raters rate each target. Consistency because systematic differences between raters are irrelevant (two-way mixed effects).

For example the paired ratings (20,40), (40,60) and (60,80) are in perfect consistency (ICC of 1.0), but with an absolute agreement of 0.6667. Consistency measures whether raters' scores are highly correlated even if they are not identical in absolute terms. That is, raters are consistent as long as their relative ratings are similar. For each of the three cases above, Shrout and Fleiss (1979) further distinguish two options; $K = 1$ or $K > 1$:

- **Single measures**: this ICC is an index for the reliability of the ratings for one, typical, single rater ($K = 1$). Measures whether raters assign the same absolute score.
- **Average measures**: this ICC is an index for the reliability of different raters averaged together ($K > 1$). This ICC is always higher than the single measures ICC.

Example of ICC(1,*k*)

ICC(1,*k*): 100 radiology centres are recruited for a study to assess computer-assisted volumetric quantification of human maxillary sinuses and each computed tomography (CT) scan is reported by the four centres closest to the originating centre. Here, the raters are different but exchangeable and thus systematic differences are relevant. Hypothetical data from six subjects are used to illustrate MedCalc output for this scenario. Each of these six scans had measurements four times from 4 of the 100 possible radiology centres (Table 3.2).

MedCalc Output

Number of subjects (n)	6	
Number of raters (k)	4	
Model	Raters for each subject are selected at random	
	One-way random effects model	
Type	Absolute agreement	
Measurements	R1	
	R2	
	R3	
	R4	
	Intraclass correlation[a]	95 % confidence interval
Single measures[b]	0.8524	0.6050–0.9744
Average measures[c]	0.9585	0.8597–0.9935

[a]The degree of absolute agreement among measurements
[b]Estimates the reliability of single ratings (ICC 1, 1)
[c]Estimates the reliability of averages of k ratings (ICC 1, 4)

Table 3.2 Data for Example ICC(1,k)

Subject	First	Second	Third	Fourth
1	Perth	Orange	Sydney	Melbourne
2	Brisbane	Hobart	Melbourne	Darwin
3	Sydney	Melbourne	Perth	Hobart
4	Darwin	Toowoomba	Orange	Sydney
5	Brisbane	Canberra	Perth	Melbourne
6	Adelaide	Townsville	Sydney	Perth
Subject	R1	R2	R3	R4
1	9	9	7	7
2	4	5	5	5
3	2	3	4	3
4	7	7	7	6
5	8	8	6	8
6	4	5	5	4

Example of ICC(2,k)

ICC(2,k): four equally experienced nurses rate six patients on a 10-point scale. Here, the raters are a random sample of nurses who will use the scale in the future and to which the assessed interobserver agreement is intended to relate. Thus, using approach 2,k would yield a suitable coefficient of agreement (Table 3.3).

MedCalc Output

Number of subjects (n)	6	
Number of raters (k)	4	
Model	The same raters for all subjects	
	Two-way model	
Type	Absolute agreement	
Measurements	Nurse1	
	Nurse2	
	Nurse3	
	Nurse4	
Absolute agreement	Intraclass correlation[a]	95 % confidence interval
Single measures[b]	0.2898	0.01879–0.7611
Average measures[c]	0.6201	0.07114–0.9272

[a]The degree of absolute agreement among measurements
[b]Estimates the reliability of single ratings (ICC 2, 1)
[c]Estimates the reliability of averages of k ratings (ICC 2, 4)

Table 3.3 Data for Example ICC(2,k)

Patients	Rater 1	Rater 2	Rater 3	Rater 4
1	9	2	5	8
2	6	1	3	2
3	8	4	6	8
4	7	1	2	6
5	10	5	6	9
6	6	2	4	7

Table 3.4 Data for Example ICC(3,k)

Patients	GM 1	GM 2	GM 3	GM 4
1	2	3	4	5
2	3	4	5	6
3	4	5	6	7
4	5	6	7	8
5	6	7	8	9
6	7	8	9	10

Example of ICC(3,k)

ICC(3,k): four different brands of glucometers are used to assess glucose levels on each blood sample (Table 3.4). The resulting intermeasure conformity concerns only these four brands and thus approach 3,k would yield a suitable coefficient of consistency.

MedCalc Output

Number of subjects (n)	6	
Number of raters (k)	4	
Model	The same raters for all subjects	
	Two-way model	
Type	Consistency	
Measurements	GM 1	
	GM 2	
	GM 3	
	GM 4	
Consistency	Intraclass correlation[a]	95 % confidence interval
Single measures[b]	1.0	
Average measures[c]	1.0	

[a]The degree of absolute agreement among measurements
[b]Estimates the reliability of single ratings (ICC 3, 1)
[c]Estimates the reliability of averages of k ratings (ICC 3, 4)

Bibliography

Agresti A (1992) Modelling patterns of agreement and disagreement. Stat Methods Med Res 1:201–218

Bland JM, Altman DG (1986) Statistical methods for assessing agreement between two methods of clinical measurement. Lancet 1:307–310

Fleiss JL, Cohen J, Everitt BS (1969) Large sample standard errors of kappa and weighted kappa. Psychol Bull 72:323–327

Krouwer JS (2008) Why Bland-Altman plots should use X, not (Y + X)/2 when X is a reference method. Stat Med 27:778–780

Lee J, Koh D, Ong CN (1989) Statistical evaluation of agreement between two methods for measuring a quantitative variable. Comput Biol Med 19:61–70

Lin L, Hedayat AS, Sinha B, Yang M (2002) Statistical methods in assessing agreement: models, issues and tools. J Am Stat Assoc 7:257–270

Sardinha LB, Lohman TG, Teixeira PJ, Guedes DP, Going SB (1998) Comparison of air displacement plethysmography with dual-energy X-ray absorptiometry and 3 field methods for estimating body composition in middle-aged men. Am J Clin Nutr 68:786–793

Shrout PE, Fleiss JL (1979) Intraclass correlations: uses in assessing rater reliability. Psychol Bull 2:420–428

Chapter 4
The Coefficient of Variation as an Index of Measurement Reliability

Orit Shechtman

Abstract This chapter focuses on the use of the coefficient of variation (CV) as an index of reliability or variability in the health sciences (medical and biological sciences) for the purpose of clinical research and clinical practice in the context of diagnostic tests, human performance tests, and biochemical laboratory assays. Before examining the use of the CV as an index of measurement reliability or variability, there is a need to define basic terms in measurement theory such as reliability, validity, and measurement error. A discussion and examples of use of the CV as a measure of reliability or variability are also provided.

Validity, Reliability and Measurement Error

Validity is the extent to which a test or an instrument measures what it is intended to measure. Reliability is the extent to which a test or an instrument measures a variable consistently. A valid test or instrument must be reliable. Reliability has many synonymous terms, including consistency, repeatability, reproducibility, stability, and precision (accuracy).

Measurement error interferes with the reproducibility of a test result or score when the measurement is repeated, which compromises reliability and affects the ability to measure change; for example, the serum level of a hormone during progression of a disease. In reality, clinical measurements are rarely completely reliable because some degree of error tends to exist in measurement instruments and in tests. Thus, an observed result is composed of a true result plus an error. Consequently, reliable clinical tests and instruments must have an acceptable amount of measurement error rather than being perfectly reliable.

O. Shechtman (✉)
Department of Occupational Therapy, College of Public Health and Health Professions, University of Florida, Gainesville, FL, USA
e-mail: Oshechtm@phhp.ufl.edu

S.A.R. Doi and G.M. Williams (eds.), *Methods of Clinical Epidemiology*, Springer Series on Epidemiology and Public Health,
DOI 10.1007/978-3-642-37131-8_4,

This uncertainty is composed of both a systematic error (bias) and a random error. Systematic error is predictable and it occurs in one direction only; random errors are unpredictable, are due to chance, are termed noise and are due to a sampling error. Sampling errors may occur for many reasons such as instrument-dependent mechanical variation, participant-dependent biological variation, examiner-dependent human error in measurement, and inconsistencies in the measurement protocol (e.g. not controlling for posture during strength measurements).

Reliability may be categorized into two types: relative reliability and absolute reliability. Relative reliability is the extent to which individuals maintain their position in a sample over repeated measurement and is assessed with correlation coefficients such as the intraclass correlation coefficient (ICC). Absolute reliability is the extent to which repeated measurements vary for an individual and is expressed by statistical methods such as the standard deviation, variance, standard error of the measurement (SEM), or the CV (See Baumgarter 1989).

The Coefficient of Variation

The CV is a statistical measure of the variability of a distribution of repeated measurements or scores (data set). A larger CV value reflects larger variability between repeated measures and therefore a smaller consistency across repeated measurements. The CV can be used as a measure of reliability because it assesses the stability of a measurement across repeated trials. Consequently, a small CV value indicates a more reliable (consistent) measurement.

The CV is a ratio between the standard deviation and the mean of a distribution of values (data set); it expresses the within-subject standard deviation as a proportion of the within-subject mean. In the special case of the distribution of the variability between repeated measurements, the CV can be defined as follows: Given that n is the number of data pairs and x_1 and x_2 are duplicate measurements for the ith pair, the formulae used to derive the within-subject SD, mean and CV are

$$SD_i = \sqrt{(x_1 - x_2)^2/2}$$

$$\bar{x}_i = (x_1 + x_2)/2$$

$$CV_i = \frac{SD_i}{\bar{x}_i}$$

$$CV(\%) = 100 \times \frac{\sum CV_i}{n}$$

As can be seen, this ratio is multiplied by 100 to give a percentage CV. The CV of a pair of repeated measurements is a unit-less or dimensionless measure because the units cancel out mathematically in the formula. An advantage of such a dimensionless measure is that it allows one to make a direct comparison between the reliability (consistency, precision) of measurements irrespective of the scale or calibration. Therefore, it permits comparison of reliability across instruments and assays (See Hopkins 2000).

The CV is a measure of relative variability (dispersion). In contrast, the common statistical measures of dispersion such as range, variance, and standard deviation are measures of the absolute variability of a data set. The CV is thus useful only if the magnitude of these measures of absolute variability change as a function of the magnitude of the data (mean). In this case, the standard deviation is inappropriate for comparing data sets of different magnitudes. On the other hand, the CV, as a measure of relative variation, expresses the standard deviation as a ratio (or a percentage) of the mean. In other words, the CV quantifies error variation relative to the mean and hence it can be used to compare the variability of data of different magnitudes. When the within-subject standard deviation is not proportional to the mean value, then there is not a constant ratio between the within-subject standard deviation and the mean and therefore, there is not one common CV. In this situation, estimating the "average" coefficient of variation is not meaningful.

Comparing Variability Across Repeated Measurements (Reliability)

The CV is used appropriately for this purpose only when the mean and standard deviation of the measurements change proportionally: the greater the mean, the greater the standard deviation. This is also called the heteroscedasticity of the data and implies that the individuals who have the highest values on a specific test also display the highest variability (standard deviation), that is, the greatest amount of measurement error. On the other hand, the data are homoscedastic when there is no relationship between the magnitude of the score and the variability.

Examples of such use include reliability of various assays, for example, blood concentrations of glucose and hormones. When glucose is measured in repeated blood sampling, the greater the mean glucose levels in the blood, the greater the SD. Also, in an enzyme-linked immunosorbent assay (ELISA), the means and standard deviation of the assay usually change proportionally, either increasing or decreasing together. The CV is a good measure of reliability in this case because it standardizes the standard deviation (by dividing it by the mean), which allows comparison of the variability of the chemical being analysed irrespective of the magnitude of its concentration.

Since the CV indicates the degree of variability in repeated tests conducted on either a particular biochemical assay or a specific person, it may be used for quality control, for example, to monitor for calibration and/or a long-term drift in repeated lipid laboratory assays of cholesterol and triglycerides. In this case, the CV is used to determine if dissimilar assay results are due to assay variability or true differences. In other words, the CV is used to establish the magnitude of differences expected in the assay due to measurement error. Similarly, the CV can be used to determine if the same person has undergone true change when two assay results are separated by an intervention such as a treatment. An intervention effect would be indicated when the two assays differ by more than expected from the variability inherent in the assay (See Reed et al. 2002).

Another paradigm is measuring change in human performance between repeated measurements over time. In this case, the CV can be used to determine if dissimilar human performance test scores are due to random variability or true differences. During rehabilitation, for example, repeated range of motion (ROM) tests are performed to quantify the outcome measures of an intervention program. The goal is to measure change in mean ROM scores before and after treatment to determine whether or not the intervention was successful in improving functional performance. Each ROM test is composed of three repeated trials (within-test sampling error) and the test is administered twice, before and after treatment (between-test change). In this case, the CV can be used to assess the reliability (precision) of ROM scores to determine if true change in mean ROM occurred due to a treatment effect. A large CV for each test (the within-test sampling error) indicates compromised reliability (precision) due to large typical error, making the documentation of true change due to an intervention effect questionable.

Comparing Variability Across Sets of Measurements on Different Scales

The CV can also be used to compare dispersion between sets of dissimilar data, for example, sets of data that are different in magnitude. As a measure of relative variability, the CV compensates for the difference in the magnitude of the data (mean). Because the magnitude of the standard deviation is a function of the magnitude of the data, dividing it by the mean to calculate the CV is an attempt to establish a measure that could be used to compare between dispersions of populations of different magnitude. The CV is appropriately used as an index of variability of different measurements only under certain conditions and when these conditions are not met, it becomes an invalid measure of reliability. Generally, the CV is validly used for this purpose only under the following conditions:

1. When describing continuous data measured on a ratio scale (not on an interval scale).
2. When comparing the variability of data sets of different magnitudes, such as data recorded in different units or on different scales.
3. When the mean difference (magnitude) and standard deviation (variability) of the measurements change (increase or decrease) proportionately. This is also called heteroscedasticity of the data.

Ratio Scale

The CV may only be used with continuous data on a ratio scale (that starts at an absolute zero) when ratio statements such as twice as big or half as much can be made. A common example from the health sciences is the measurement of blood concentrations of drugs. The CV is not valid for interval scales, where the zero value is arbitrary and ratios between numbers are not meaningful. An example of an interval scale would be temperature recorded in degrees Celsius or Fahrenheit. Also, blood pressure, heart rate and weight do not have an absolute zero in living persons. According to Allison (1993), without an absolute zero value, the mean, which is the denominator of the CV, becomes an arbitrary value, which renders the CV invalid.

Different Magnitudes

The range, variance and standard deviation cannot be used in the case of different magnitudes as they are measures of absolute variability and thus their magnitude depends on the scale of the data. Since the CV is a dimensionless (unitless) statistical measure, it can be validly used to compare the variability of data recorded in different units or magnitudes.

A simple example of using different scales is making a comparison between organ size in various species, such as the nose length of elephants and mice. The nose lengths of these two species are very different, that is, several feet for elephants versus a fraction of an inch for mice. Thus, the absolute variability, for example, standard deviation, can be expected to be much larger for a distribution of elephant nose lengths than for a distribution of mouse nose lengths. Shechtman (2000) gave a numerical example of a distribution of nose lengths (expressed as mean $\pm$ standard deviation), assuming elephant nose length to be 72 ± 6 inches and mouse nose length to be 0.07 ± 0.005 inches. To correctly compare the variability between the two species, the mean nose length has to be taken into account because the absolute variability (standard deviation) is much larger for the elephant than for the mouse nose length (6 inches vs. 0.005 inches, respectively). On the other hand, the relative variation as expressed by the CV ($CV = SD/mean \times 100$) is 8.3 % and 7.1 % for the elephant and mouse nose lengths, respectively. Consequently, the CV, which expresses the standard deviation as a percent of the mean, is a more valid comparison of variability between the two distributions.

Heteroscedasticity

The CV is not a valid measure of reliability with homoscedastic data because it becomes inflated. For example, when the mean of one distribution is smaller compared with the mean of another distribution, but the standard deviations of the two distributions are similar, the CV becomes inflated. With homoscedastic data sets, the premise that a larger CV indicates more variability in the data is incorrect. When calculating the CV, the standard deviation is the nominator and the mean is the denominator (CV = SD/mean), so the CV becomes larger due to either a larger standard deviation or a smaller mean. If an experiment is being conducted to compare grip strength between an injured and uninjured hand, because grip strength is reduced in one hand due to hand injury, the larger CV is due to the decrease in the mean and not as a result of a true increase in variability, because the standard deviation is similar for the injured and uninjured hands. Similarly, in a study comparing the variability of ROM of different joints, Bovens et al. (1990) state that the CV of ankle eversion is large because the mean ROM is small. Thus, the use of the CV is valid only when comparing distributions in which the mean changes proportionally to the standard deviation (variability) of measurements.

Examples

Examples of variability of human performance measurements include determining the variability of hand-held dynamometry for measuring muscle strength. Because various muscle groups have different magnitudes of strength (mean force ranging from 4.8 to 13.7 kg), the CV can be used to determine the variability of hand-held dynamometry for measuring strength across the different muscle groups (see Wang et al. 2002). Another example is comparing the variability of excursion of back extension in centimetres versus degrees when assessing spinal flexibility with different instruments. An additional example is comparing the variability in blood cholesterol concentrations with the variability in blood vessel diameter. In a different study, Mannerkorpi and Co-authors (1999) assessed the variability of seven different tests in a test battery for functional limitations in fibromyalgia syndrome using the CV. In a study with a narrower scope Montgomery and Gardner (1998) used the CV to determine test–retest reliability of different measures derived from the 6-min walk test, that is, the number of steps taken versus the total walking distance. In the above examples, it is valid to use the CV because it measures the relative spread of the data and it adjusts the scales that are represented in different units.

In the case of large, multicenter, epidemiological studies in which multiple tests and multiple examiners are necessary, consistency and precision in data collection are critical, and the CV may be used to assess the variability (reliability) of the various data sets. For example, Klipstein-Gnobusch et al. (1997) compared the variability of data sets collected by 17 different raters recording multiple anthropometric measurements using the CV and other measures of reliability. The authors used the CV to assess both inter-rater and intra-rater measurement error and to describe the precision of data collection for both the different raters and the various tests. As expected, the CVs were smaller for the more objective tests, such as body weight and height, compared with the more subjective circumference and skin-fold tests. The smaller CVs indicated smaller variability for the objective tests as well as a smaller intra-rater than inter-rater variability. In this study, the CV was used to express the measurement error inherent in these tests as a proportion of the mean, to reflect the variability of the scores as independent of the between-subject variance.

Invalid Use of the CV for Comparing Variability Across Data Sets

Using the CV for comparing variability where the three conditions set out are not met can result in invalid comparisons. The measure of sincerity of effort is a good example of such invalid use. The CV is widely used to determine the sincerity of effort of grip strength measurements. Grip strength scores are often used to determine the extent of an injured worker's disability and the amount of the subsequent financial compensation. Grip strength measurements, however, are objective, reliable, and valid only when an individual exerts a sincere maximal effort. Intentional exertion of an insincere, feigned, and submaximal effort could lead to unnecessary medical procedures and lack of response to intervention, and thus could lead to rising health care costs.

The rationale for using a measure of variability to assess sincerity of effort is based on the motor unit recruitment model, which suggests that repeated maximal efforts are expected to be more consistent than repeated submaximal efforts (Kroemer and Marras 1980). According to this model, it should be easy to achieve consistent repetitions of maximal muscular contractions because simple motor control is required for maximal firing frequency and maximal motor unit recruitment. On the other hand, consistent repetitions of submaximal efforts should be more difficult to accomplish because higher levels of motor control are necessary for grading muscular contraction, including constant corrections of motor signals based on delicate proprioceptive feedback.

As a test of sincerity of effort, the CV is used to compare the variability of at least three repeated grip strength trials between the injured hand and the uninjured hand. The premise is that a sincere maximal effort should be more consistent than an insincere submaximal effort regardless of hand injury, and therefore repeated maximal efforts should have smaller CVs. Several studies investigating the validity of using the CV as a measure of sincerity of effort suggest that the reason for the larger CV of submaximal efforts is the smaller mean grip force and not a true increase in the variability of the data set (Shechtman 2000).

For the CV to validly indicate the variability of repeated grip strength trials, the data set must be heteroscedastic, which means that a stronger person (with greater mean grip strength) must have greater variability (standard deviation) than a weaker person. In other words, the stronger person must be less consistent in repeating strength trials than the weaker person. Otherwise, the CV would be inflated for the weaker person as previously discussed.

Shechtman (1999) presented case studies of three patients who were tested for grip strength before carpal tunnel release surgery and then at 6, 13, and 26 weeks after surgery, each performing three repeated grip trials at each time point. The study clearly demonstrated homoscedastic data sets: although the mean grip strength for each participant decreased after surgery and slowly recovered with time, the standard deviation of the three grip trials remained essentially the same.

To demonstrate the problem of using the CV with homoscedastic data, consider the data sets presented in Figs. 4.1, 4.2, and 4.3. The data in these figures show no relationship between the mean and standard deviation, as the mean (Fig. 4.1) changes without a corresponding proportional change in the standard deviation (Fig. 4.2). As a consequence, the CV of the weaker patient (patient 2) is inflated, especially at 6 and 13 weeks after surgery (Fig. 4.3). These large CVs are due to a smaller mean rather than to a true increase in variability (standard deviation). The lack of proportional change between the means and standard deviations is demonstrated for both within-subject and between-subject grip strength scores, as the standard deviation values are similar throughout the data sets (Fig. 4.2).

In a study with a larger sample size, Shechtman (2001a) again demonstrated that repeated grip strength trials produce homoscedastic data sets. Healthy participants ($n = 146$) performed five repeated grip strength trials twice, once exerting maximal effort and once exerting submaximal effort. The data showed that larger mean torques did not yield larger standard deviations: the mean torque of the submaximal efforts was 50 % smaller than that of the maximal efforts and the standard deviation of the submaximal efforts was 5 % larger than that of the maximal efforts. This lack of relationship between the means and standard deviations violates the statistical principle of heteroscedasticity of data, resulting in biased outcomes expressed as inflated CVs for the submaximal efforts. The author concluded that the CV is not an appropriate measure of variability for determining sincerity of effort of grip strength because the increased CVs associated with submaximal efforts were not due to a true increase in variability. The increased CV values were due to a decrease in the mean, and thus did not indicate a lack of consistency or insincerity. When CV values are inflated, a person may be wrongly accused of feigning an effort. Mistakenly labelling a sincere effort as insincere may result in inappropriate diagnosis and treatment, reduced financial compensation and job loss, all of which have a negative impact on the individual patient.

An additional problem that may arise when using the CV is that the cut-off value used to indicate what is an acceptable variability in a data set is often chosen arbitrarily. In the specific case of sincerity of effort, the cut-off value found in the related literature varies greatly; it ranges from 7.5 % to 20 %. The most commonly used CV cut-off value is 15 %. Instead of arbitrarily choosing a cut-off value, Shechtman (2001b) calculated sensitivity and specificity values for the entire range

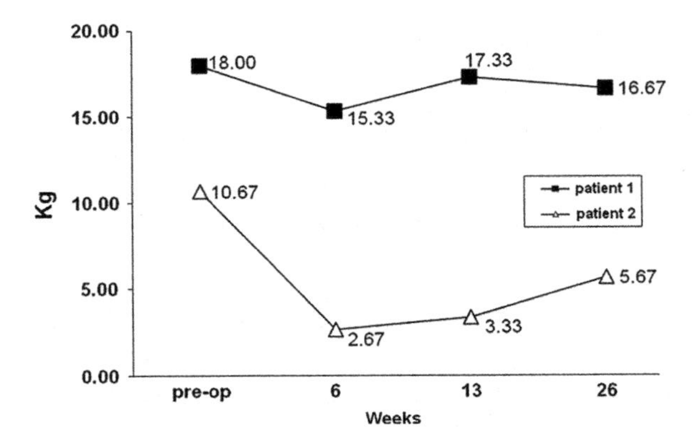

Fig. 4.1 Mean grip strength (kg) of three repeated trials performed by two patients before carpal tunnel surgery and at 6, 13, and 26 weeks after surgery

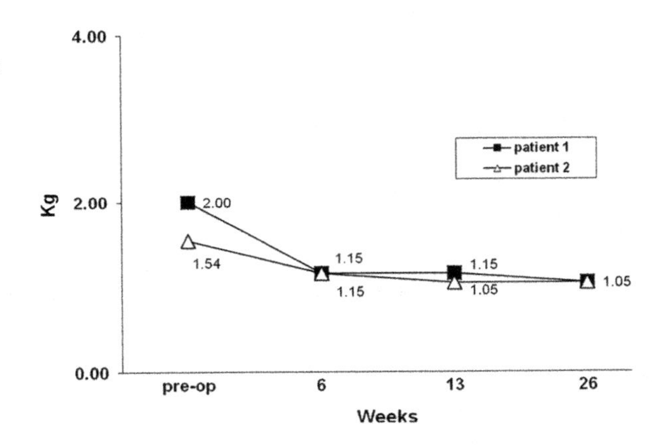

Fig. 4.2 Standard deviation of three repeated grip strength trials (kg) performed by two patients before carpal tunnel surgery and at 6, 13, and 26 weeks after surgery

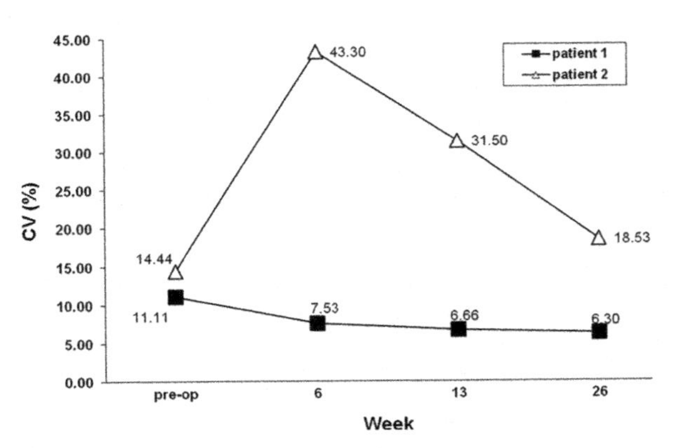

Fig. 4.3 Coefficients of variation of three repeated grip strength trials (%) performed by two patients before carpal tunnel surgery and at 6, 13, and 26 weeks post surgery

of CV cut-off values. The author concluded that the CV lacked a sufficient combination of sensitivity and specificity values necessary for detecting sincerity of effort of grip strength in clinical practice.

Conclusions

The advantages of the CV as a measure of data reliability are that it is simple to calculate and it is dimensionless (unitless). As a unitless measure, the CV is used to compare numerical distributions that are measured on different scales because it does not vary with changes in the magnitude of a measurement. The CV permits direct comparison of variability of different measurements without the need to take into account either calibration or the magnitude of scores. Therefore, the CV allows comparison of variability across various instruments, testers (raters), and study samples (populations). In addition, the CV can be used to evaluate whether or not differences between two measurements are due to random variation and thus enabling true changes to be tracked in clinical settings. Although the existence of statistical tests to measure significant differences between CVs was disputed, Tian (2005) proposed a statistical method for comparing between CVs.

The limitations of the CV as an index of measurement reliability stem from using it incorrectly by violating the three conditions of valid use. The CV is not valid when used with data sets that are: (1) not recorded as ratio scales; (2) not of different magnitudes or units; or (3) not heteroscedastic (the mean and standard deviation do not change proportionally). Some authors also caution against using an arbitrarily chosen cut-off value for the CV when determining the variability of data sets.

Bibliography

Allison DB (1993) Limitations of coefficient of variation as index of measurement reliability. Nutrition 9:559–561

Ashfor RF, Nagelburg S, Adkins R (1996) Sensitivity of the Jamar dynamometer in detecting submaximal grip effort. J Hand Surg 21A:402–405

Atkinson G, Nevill AM (1998) Statistical methods for assessing measurement error (reliability) in variables relevant to sports medicine. Sports Med 26:217–238

Baumgarter TA (1989) Norm-referenced measurement: reliability. In: Safrit MJ, Wood TM (eds) Measurement concepts in physical education and exercise science. Human Kinetics, Champaign, pp 45–72

Bohannon RW (1987) Differentiation of maximal from submaximal static elbow flexor efforts by measurement variability. Am J Phys Med 66:213–218

Bohannon RW (1997) Hand-held dynamometry: factors influencing reliability and validity. Clin Rehabil 11:263–264

Bohannon RW (1999) Intertester reliability of hand-held dynamometry: a concise summary of published research. Percept Mot Skills 88(3 Pt 1):899–902

Bovens AM, van Baak MA, Vrencken JG, Wijnen JA, Verstappen FT (1990) Variability and reliability of joint measurements. Am J Sports Med 18:58–63

Bruton A, Conway JH, Holgate ST (2000) Reliability: what is it, and how is it measured? Physiotherapy 86:94–99

Dvir Z (1999) Coefficient of variation in maximal and feigned static and dynamic grip efforts. Am J Phys Med Rehabil 78:216–221

Goto R, Mascie-Taylor CG (2007) Precision of measurement as a component of human variation. J Physiol Anthropol 26:253–256
Hamilton Fairfax A, Balnave R, Adams RD (1995) Variability of grip strength during isometric contraction. Ergonomics 38:1819–1830
Hamilton Fairfax A, Balnave R, Adams R (1997) Review of sincerity of effort testing. Saf Sci 25:237–245
Hopkins WG (2000) Measures of reliability in sports medicine and science. Sports Med 30:1–15
Klipstein-Grobusch K, Georg T, Boeing H (1997) Interviewer variability in anthropometric measurements and estimates of body composition. Int J Epidemiol 26:S174–S180
Kroemer KH, Marras WS (1980) Towards an objective assessment of the "maximal voluntary contraction" component in routine muscle strength measurements. Eur J Appl Physiol Occup Physiol 45:1–9
Lachin JM (2004) The role of measurement reliability in clinical trials. Clin Trials 1:553–566
Lechner DE, Bradbury SF, Bradley LA (1998) Detecting sincerity of effort: a summary of methods and approaches. Phys Ther 78:867–888
Mannerkorpi K, Svantesson U, Carlsson J, Ekdahl C (1999) Tests of functional limitations in fibromyalgia syndrome: a reliability study. Arthritis Care Res 12:193–199
Mathiowetz V, Weber K, Volland G, Kashman N (1984) Reliability and validity of grip and pinch strength evaluations. J Hand Surg 9:222–226
Mitterhauser MD, Muse VL, Dellon AL, Jetzer TC (1997) Detection of submaximal effort with computer-assisted grip strength measurements. J Occup Environ Med 39:1220–1227
Montgomery PS, Gardner AW (1998) The clinical utility of a six-minute walk test in peripheral arterial occlusive disease patients. J Am Geriatr Soc 46:706–711
Muggeo M, Verlato G, Bonora E, Zoppini G, Corbellini M, de Marco R (1997) Long-term instability of fasting plasma glucose, a novel predictor of cardiovascular mortality in elderly patients with non-insulin-dependent diabetes mellitus: the Verona Diabetes Study. Circulation 96:1750–1754
Portney LG, Watkins MP (2000) Foundations of clinical research: applications to practice, 2nd edn. Appleton & Lange, Norwalk
Reed GF, Lynn F, Meade BD (2002) Use of coefficient of variation in assessing variability of quantitative assays. Clin Diagn Lab Immunol 9:1235–1239
Robinson ME, MacMillan M, O'Connor P, Fuller A, Cassisi JE (1991) Reproducibility of maximal versus submaximal efforts in an isometric lumbar extension task. J Spinal Disord 4:444–448
Robinson ME, Geisser ME, Hanson CS, O'Connor PD (1993) Detecting submaximal efforts in grip strength testing with the coefficient of variation. J Occup Rehabil 3:45–50
Shechtman O (1999) Is the coefficient of variation a valid measure for detecting sincerity of effort of grip strength? Work 13:163–169
Shechtman O (2000) Using the coefficient of variation to detect sincerity of effort of grip strength: a literature review. J Hand Ther 13:25–32
Shechtman O (2001a) The coefficient of variation as a measure of sincerity of effort of grip strength, Part I: the statistical principle. J Hand Ther 14:180–187
Shechtman O (2001b) The coefficient of variation as a measure of sincerity of effort of grip strength, Part II: sensitivity and specificity. J Hand Ther 14:188–194
Shechtman O, Anton SD, Kanasky WF Jr, Robinson ME (2006) The use of the coefficient of variation in detecting sincerity of effort: a meta-analysis. Work 26:335–341
Simonsen JC (1995) Coefficient of variation as a measure of subject effort. Arch Phys Med Rehabil 76:516–520
Streiner DL, Norman GR (1989) Health measurement scales: a practical guide to their development and use. Oxford University Press, New York
Tian L (2005) Inferences on the common coefficient of variation. Stat Med 24:2213–2220
Wang CY, Olson SL, Protas EJ (2002) Test-retest strength reliability: hand-held dynamometry in community-dwelling elderly fallers. Arch Phys Med Rehabil 83:811–815

Part II
Diagnostic Tests

Chapter 5
Using and Interpreting Diagnostic Tests with Dichotomous or Polychotomous Results

Cristian Baicus

Abstract In order to use and interpret diagnostic tests, we need to know the operating characteristics of the test (sensitivity, specificity, predictive values and likelihood ratios), and the prevalence of the disease. Sensitivity is the probability that a sick person tests positive, and a very sensitive test is useful when negative, to rule out the disease. Specificity is the probability of a healthy person to test negative, and a very specific test is useful when positive, to rule in the diagnosis. The positive predictive value is the probability that a person whose test is positive has the disease; the negative predictive value is the probability that a person whose test is negative is healthy. The predictive values depend on the prevalence of disease (pretest probability) so that, for the same predictive values, the probability of disease is higher in settings with a higher prevalence of disease than in those with a lower prevalence (Bayes' rule). The likelihood ratios of a test, which may be calculated from its sensitivity and specificity, are stable for different prevalences, can deal with both dichotomous and polychotomous (multilevel) test results, and can also be used to calculate the post-test probability. For a valid diagnostic study, consider the standards for reporting of diagnostic accuracy (QUADAS) criteria.

Sensitivity and Specificity

Disease is a categorical variable that can have two values (the patient either has or has not the disease), therefore it is a dichotomous variable. The diagnostic test can also be a dichotomous variable (e.g. imaging) or an interval variable (e.g. laboratory testing

C. Baicus (✉)
Clinical Epidemiology Unit, Bucharest and Carol Davila University of Medicine and Pharmacy, Bucharest, Romania
e-mail: cbaicus@clicknet.ro

S.A.R. Doi and G.M. Williams (eds.), *Methods of Clinical Epidemiology*, Springer Series on Epidemiology and Public Health,
DOI 10.1007/978-3-642-37131-8_5,

Table 5.1 A 2 × 2 contingency table with the four possibilities concerning the results of a diagnostic test

Test	Disease (gold standard)	
	Present	Absent
Positive	TP	FP
Negative	FP	TN

FN false-negative, *FP* false-positive, *TN* true-negative, *TP* true-positive

for haemoglobin, serum glucose, prostate-specific antigen, etc.), which can be transformed into dichotomous variables by using a cut-off level.[1]

Therefore, when comparing two dichotomous variables, we use a 2 by 2 contingency table (Table 5.1), with the outcome variable (disease) in the columns, and the predictive variable (result of the diagnostic test) in the rows. There are four possible situations: the patient has the disease and the test is positive (true-positive); the patient has the disease but the test is negative (false-negative); the patient does not have the disease and the test is negative (true-negative); and the patient does not have the disease but the test is positive (false-positive).

The accuracy of a test is the proportion of correct diagnoses (true-positives and true-negatives) among all tested patients (Table 5.2).

The sensitivity (Se) of a test is the probability of a positive test when the person is sick, or the proportion with a positive test among all people with the disease (people with the disease who have a positive test divided by all the people with the disease; positive in disease, PID); therefore, the sensitivity is calculated on the people who have the disease.

The sensitivity represents the power of a test to discover the disease. The more sensitive the test, the less the risk for patients with the disease going undiscovered (1 − Se = the false-negative rate: the higher the sensitivity, the less the false-negatives). A very sensitive test (close to 100 %) is most useful when it is negative. The proportion of false-negatives is very small, therefore we can rule out the disease (S_nOUT) (Sackett et al. 1991).

The specificity (Sp) is the probability of a healthy individual testing negative, or the proportion of people with a negative test among the healthy persons. Therefore, the specificity is calculated in healthy people.

A very specific test (close to 100 %) is useful when it is positive, because it rules in the disease (S_pIN); specificity is inversely correlated with the false-positive rate (1 − Sp = false-positive rate).

The ideal test is sensitive and specific at the same time, because it is useful to rule in the disease when it is positive and rule out the disease when it is negative.

Because sensitivity and specificity are calculated on sick people and healthy people, respectively, they do not depend on the prevalence of the disease; therefore these characteristics remain constant in any setting.[2]

[1] For the evaluation of tests with an interval scale, the choice of a cut-off level and the utilization of tests with more than two levels, see Chap. 6.

[2] At least theoretically. In fact, the sensitivity of a test is higher at the tertiary care level because the patients have more advanced disease than in primary care.

Table 5.2 A 2 × 2 contingency table for the assessment of a diagnostic test

	Disease		
Diagnostic test	Present	Absent	Total
Positive	*a*	*b*	$a + b$
Negative	*c*	*d*	$c + d$
	$a + c$	$b + d$	$a + b + c + d$

a true-positives, *b* false-positives, *c* false-negatives, *d* true-negatives
Accuracy = $(a + d)/(a + b + c + d)$
Sensitivity (Se) = $a/(a + c)$
Specificity (Sp) = $d/(b + d)$
Pretest probability = prevalence of disease = $(a + c)/(a + b + c + d)$
Positive predictive value (PPV) = $a/(a + b)$
Negative predictive value (NPV) = $d/(c + d)$

The Predictive Values of Tests

The sensitivity and specificity, although important parameters of the diagnostic tests, do not help us in clinical practice, except when they are close to 100 % and we apply the S_nOUT and S_pIN rules. This happens because their definition is the reverse of the diagnostic approach. In medical practice, we do not know if the patient is sick or healthy; we are trying to find out the probability of the sick or healthy patient testing positive or negative respectively (the definitions of sensitivity and specificity). Conversely, we are applying a diagnostic test and we want to know, depending on its result, what is the probability of disease. This is the definition of predictive values.

The positive predictive value (PPV) is the probability that a person with a positive test is a true-positive. The negative predictive value (NPV) is the probability that a person with a negative test does not have the disease (Last 2001).

We can compute the predictive values using the same 2 × 2 contingency table (Table 5.2). The predictive values (positive = $a/(a + b)$ and negative = $d/(c + d)$), unlike sensitivity and specificity, are calculated on the horizontal lines, therefore they depend on the prevalence of the disease (= $(a + c)/(a + b + c + d)$). This means that, while sensitivity and specificity are relatively constant in different setting with different prevalences, the predictive values are highly dependent on prevalence. This is the Bayes' theorem: the post-test (or posterior) probability depends on the pretest (or prior) probability; or the probability of disease given this diagnostic test depends not only on how good the test is (sensitivity and specificity), but also on the prevalence of the disease in the tested population. This is summarized by the following formulas:

$$\text{PPV} = \frac{\text{Se} \times \text{Prevalence}}{\text{Se} \times \text{Prevalence} + (1 - \text{Sp}) \times (1 - \text{Prevalence})}$$

$$\text{NPV} = \frac{\text{Sp} \times (1 - \text{Prevalence})}{\text{Sp} \times (1 - \text{Prevalence}) + (1 - \text{Se}) \times (\text{Prevalence})}$$

Table 5.3 Post-test probability for a very good diagnostic test (Se = Sp = 95 %)

Prev	99	95	90	80	70	60	50	40	30	20	10	5	1
PPV	99.9	99.7	99.4	99	98	97	95	93	89	83	68	50	16
NPV	16	50	68	83	89	93	95	97	98	99	99.4	99.7	99.9
NTP	84	50	32	17	11	7	5	3	2	1	0.6	0.3	0.1

Prev prevalence = pretest probability, *PPV* positive predictive value = post-test probability (positive test), *NPV* negative predictive value, *NTP* post negative test probability = 100 − NPV

Table 5.4 Post-test probabilities for a diagnostic test with Se = Sp = 90 % (diagnostic criteria for giant cell arteritis)

Prev	99	95	90	80	70	60	50	40	30	20	10	5	1
PPV	99.8	99.4	98.7	97	95	90	90	86	79	69	50	32	8.3
NPV	8.3	32	50	69	79	86	90	93	95	97	98.7	99.4	99.9
NTP	81.7	58	50	31	21	14	10	7	5	3	1.3	0.6	0.1

Prev prevalence = pretest probability, *PPV* positive predictive value = post-test probability (positive test), *NPV* negative predictive value, *NTP* post negative test probability = 100 − NPV

Therefore, a low prevalence determines the decrease in the PPV and the increase in the NPV; a high prevalence has the opposite effect.

Table 5.3 presents how a very good test (with Se = Sp = 95 %) changes the post-test probability, for different prevalences of a disease. For very high (95–99 %) or very low (1–5 %) prior probabilities, the test is useless, because the shift from pre-test to post-test probability is not large enough to change the classification of the patient from having the disease to not having the disease and vice versa. If the patient has a 95 % pretest probability of having a disease, we know he has the disease, and other tests are not necessary; for a 5 % pretest probability, we already know he does not have the disease, and other tests are not necessary. A test is useful when applied to patients with average prior probabilities (with the largest shift of probability for a 50 % prior probability: from 50 % to 95 % if the test is positive, or from 50 % to 5 % if the test is negative).

Unfortunately, the diagnostic tests are rarely so good, and for a weaker test (smaller Se and Sp), the probability shifts determined by the test are smaller too (Table 5.4).

Calculating the Post-Test Probability (PPV) of a Given Test for Variable Prevalences

According to Bayes' theorem, we need the pretest probability in order to calculate the post-test probablility (= PPV). How can we find this?

We can never know it for sure, but it can be estimated. For primary care, it is equal to the prevalence of the disease in the local population, and if we do not know this, we can extrapolate it from studies in similar populations. We can find these prevalences in the literature by searching for descriptive studies documenting the

Table 5.5 Calculation of the predictive values for a test knowing the prevalence, the sensitivity and the specificity (the back calculation method)

	Disease		
Diagnostic test	Present	Absent	Total
Positive	a	b	$a + b$
Negative	c	d	$c + d$
	$a + c$	$b + d$	$a + b + c + d$

a true-positives, b false-positives, c false-negatives, d true-negatives
We know that the pretest probability (prevalence) = $(a + c)/(a + b + c + d) = 5$ %. Se = $a/(a + c) = 0.95$; Sp = $d/(b + d) = 0.90$. We do not know: PPV = $a/(a + b)$; NPV = $d/(c + d)$

pretest probabilities for diseases with underlying signs and symptoms similar to those presented by our patient. Another way to find the prevalence is to look at the prevalence of the disease among the patients in our practice or service in recent years.

Once the prevalence is established, we need to find in the literature the parameters (Se and Sp) for the diagnostic test we are interested in. If they were calculated in an environment similar to that in which we work (the same disease prevalence), then we can use the predictive values calculated in that article. Otherwise, relying on the fact that Se and Sp remain constant regardless of the prevalence, we can calculate the post-test probability (or PPV) by the back calculation method, using a 2 × 2 table again (see Table 5.2). If a test has Se = 95 % and Sp = 90 %, and the prevalence of the disease is 5 % in our setting, we can calculate the PPV and NPV for our patient after he tests positive or negative, respectively.

First, we give an arbitrary value to the cell with the total number of patients in the lower right corner (e.g. $a + b + c + d = 1{,}000$; Table 5.5). Knowing the prevalence of the disease, which is $(a + c)/(a + b + c + d)$, we find the value for $a + c$. For a prevalence of 5 % in our setting, there will be 50 patients with the disease ($a + c$) and 950 people without the disease (1,000 − 50), and we can complete all the cells at the bottom of the table.

Then, starting from the known value of the Se = 95 %, we calculate the values in the column for the patients who have the disease: Se = $0.95 = a/(a + c) = a/50$, which means $a = 0.95 \times 50 = 47.5$, rounded to 47 true-positives; therefore the number of false-negatives c is $50 - 47 = 3$, and we have the first column completed (Table 5.6).

Knowing that Sp = 0.90, we calculate the number of true-negatives = d = $950 \times 0.90 = 855$, and the number of false-positives = $950 - 855 = 95$, and we have the contingency table completed. Therefore we can proceed to the calculation of the predictive values: PPV = 47/142 = 0.33, and the NPV = 855/858 = 0.997. A simpler method to calculate the predictive values according to prevalence is by using likelihood ratios (see later).

Table 5.6 The back calculation of the values in the contingency table to adapt the predictive values of a test to a different prevalence

	Disease		
Diagnostic test	Present	Absent	Total
Positive	47	95	142
Negative	3	855	858
	50	950	1,000

PPV = 0.33; NPV = 0.997

Table 5.7 An example of a biochemical test used to diagnose disease X

Serum levels for biochemical test	Disease present	Disease absent	Serum levels for biochemical test	Disease present	Disease absent
≤18	47	2	≤45	70	15
19–45	23	13		70/85 = 0.824	15/150 = 0.1
				Sensitivity	1 − specificity
46–100	7	27	>45	15	135
>100	8	108		15/85 = 0.176	135/150 = 0.9
				1 − sensitivity	Specificity
Totals	85	150	Totals	85	150

Dichotomizing Categories of a Diagnostic Test

Dichotomizing leads to loss of information so the true-positive and false-positive rates tend to become closer together. Dichotomization thus increases the risk of a positive result being a false-positive. Individuals close to but on opposite sides of the cut-off point are characterized as being very different rather than very similar. Using two groups conceals any non-linearity in the relationship between the variable and outcome. Also, if dichotomization is used, where should the cut-off point be? For a few variables there are recognized cut-off points, such as body mass index (BMI, calculated as weight in kilograms divided by the square of height in meters) > 25 kg/m^2 defines overweight. In the absence of a prior cut-off point (e.g. for disease X in Table 5.7), this is an arbitrary process. Therefore, in such situations, it is preferable to compute the sensitivity and specificity for each category (used as the decision threshold) or, better still, to compute the likelihood ratios (see later).

Multiple Testing

As most diagnostic tests are far from being perfect, a single test is generally not enough. For this reason, physicians use multiple diagnostic tests, administered either in parallel (simultaneously) or serially (sequentially).

In parallel testing, the tests are done simultaneously. The test battery is considered negative when all the tests are negative, and positive when at least one of the

Table 5.8 Anaemia, erythrocyte sedimentation rate (ESR) and involuntary weight loss as diagnostic tests for cancer (Baicus et al. 1999): parallel testing increases the sensitivity; serial testing increases the specificity

Test	Se (CI)	Sp (CI)
Anaemia	37 (36–39)	92 (91–93)
ESR	52 (51–54)	89 (88–90)
Weight loss	46 (45–48)	94 (93–94)
Parallel testing	87 (86–88)	79 (78–81)
Serial testing	9 (9–10)	99.6 (99–100)

CI confidence interval

tests is positive. Using parallel testing, there is a gain in sensitivity at the expense of a loss in specificity compared with either of the tests used alone. For example, we could say that a patient with arthritis has systemic lupus erythematosus (SLE) if she has any one of malar rash, nephrotic syndrome, thrombocytopenia, pleural effusion or antinuclear antibodies (ANA), and we can decide she does not have SLE if she has none of these symptoms. We will miss very few patients with lupus, but probably we will have more false-positives.

By applying the same tests serially, we decide that the patient has SLE only if she has, concomitantly, malar rash, nephrotic syndrome, thrombocytopenia, pleural effusion and ANA. We rule out the disease if the patient does not have all of these symptoms. Using this method, we increase the specificity (we are sure that a patient with all these symptoms has SLE) but lose sensitivity (a lot of patients will not have all the symptoms so there will be many false-negatives).

Most of the time, tests are done sequentially, beginning with the most specific (if the tests pose equal risk & cost to the patient) or, in daily practice, from the simplest to the most expensive or invasive: history, clinical examination, and laboratory tests. The tests are done in parallel when a rapid assessment is necessary, for example, in an emergency or in hospitalized patients.

Table 5.8 presents data on anaemia, erythrocyte sedimentation rate and weight loss as diagnostic tests for cancer, and the variations in sensitivity and specificity that appear with the parallel or serial use of these tests.

Parallel and serial testing do not use the value of every test in the battery of tests, but only the battery as a whole: for parallel testing, if only one of the tests is positive or all are positive, we consider the battery positive and the disease present (*"believe the positive"*); for serial testing, if one test is negative or all are negative, we consider the battery negative and the disease absent (*"believe the negative"*).

For more complex evaluations where the value of every test is taken into account, multivariable analysis (logistic regression) can be used, and diagnostic or prognostic models and scores or clinical prediction rules can be developed.

The Gold (Reference) Standard

To estimate classification accuracy using diagnostic tests, the disease status for each patient has to be measured without error (i.e. the end point has to be defined without uncertainty; however, this is sometimes not the case). The presence or absence of the disease state is defined according to some, sometimes arbitrarily selected, gold standard. Because, in reality, there is no perfect test, some people call this the reference standard. Thus, the true disease status is often inferred from the reference standard or the gold standard. The reference standard may be available from clinical follow-up, surgical verification, biopsy and autopsy. For certain diseases, there is no single test that can constitute the gold standard, and in this case, the investigators build one relying on multiple tests or on clinical evolution. In the latter situation, a set of criteria defining the outcome has to be set before the start of the study, and the outcome should be evaluated by an independent committee of experts or in particular situations by the use of results of multiple imperfect tests (referred to as a composite reference standard). The nature of the reference standard can itself be a cause for debate.

The reference standard is generally the most expensive or invasive of all available tests, and this is the reason why researchers try to find new, cheaper or less invasive tests, with the hope of demonstrating that are comparable with the reference standard, in diagnostic research studies. In the event that a new test is better than the reference standard, the usual diagnostic study design will not be able to reveal this, because, for example, diseased individuals correctly detected only by the new test will be falsely labeled as false-positives.

Diagnostic Test Research: Study Design and Validity Criteria

The classic study design is a cross-sectional study in which both the diagnostic test and the reference standard are applied in every patient, concomitantly and independently. If the test is a prognostic one (or the reference standard is clinical evolution), then the design is a cohort study.

The case–control design, in which people with and without the disease are sampled separately, usually compares people with clear disease with healthy people, and thus the results of the study are over optimistic concerning the discriminative power of the diagnostic test because the test is not assessed on a clinically realistic spectrum of patients. For that reason, case–control diagnostic studies are less valid, and they should be reserved for rare diseases or as a first stage diagnostic study, knowing that if low sensitivity and/or specificity are found, it is not worthwhile continuing with a more valid type of study.

In order to be valid, a diagnostic test study has to meet the following criteria:

1. There must be an independent, blind comparison with a reasonable reference standard. Independent means that there must not be common items in the diagnostic test and the reference standard, and the results of the reference standard must be independent of the results of the evaluated test (incorporation bias). Blind means that the investigators who interpret the test must not be aware of the results of the reference standard, and those who interpret the reference standard must not be aware of the results of the test.

2. The reference standard must be applied to all the patients included in the study, even if the evaluated test is negative, in order to detect the false-negatives (for invasive tests, there are alternatives such as clinical follow-up).
3. The studied test must be evaluated for reproducibility (see Chaps. 1 and 2 on the kappa statistic and clinical agreement).
4. The study population must be similar to the population to be tested (in order to avoid spectrum bias). For this purpose, a cross-sectional design must be used, and all consecutive patients suspected of the disease for which the test is evaluated must be included. This kind of study (unlike the case–control approach) will reproduce the true prevalence of the disease in that setting, and the predictive values will be true (for that setting).
5. The sample size must be calculated (see Chap. 7) in order to have acceptable precision for the sensitivity, specificity and likelihood ratios without large confidence intervals, which would make the results of the study uninterpretable (no wider than $\pm$ 10 %).

Generally, for a valid diagnostic study, it is better to consider the standards for reporting of diagnostic accuracy (STARD) initiative (http://www.stard-statement.org), which was published in 2003 in several leading medical journals.

But diagnostic testing is only an intermediate end point, and one cannot be sure that by using a better diagnostic test, the patient will live longer or better. The answer to this question is given only by randomized controlled studies, in which patients are randomized to receive either the standard diagnostic tests or the new diagnostic tests, and then they are treated according to the results of the diagnostic work-up. Using this kind of design, if the new test is better than the reference standard used, this can be found.

At the end, even if the diagnostic model fulfills all the validity criteria, it can only be trusted after validation in a second sample (and ideally in different settings). Until this kind of validation (external), which requires a new sample and may be a new study, has been done, one could be content with an internal validation, achieved by simulation of a second data set and using the statistical techniques of split sampling, jackknifing or bootstrapping.

The Likelihood Ratio

A likelihood ratio (LR) is the ratio of the probability of a positive test (LR+) or negative test (LR−) in the diseased over the healthy. Thus, an $LR > 1$ means that the test result is more likely to result in disease; $LR < 1$ means the test result is more likely to result in no disease; $LR = 1$ means that the test result is equally likely in the diseased and non-diseased. LRs describe how much the odds change after applying the results of a test.

$$LR^+ = Se/(1 - Sp), \quad LR^- = (1 - Se)/Sp$$

The LRs tell us by how much a given diagnostic test result will increase or decrease the pretest assessment of the chance of the target disorder by the physician. Thus:

- An LR of 1 means that the post-test probability is exactly the same as the pretest probability.
- LR $>$ 1.0 increases the probability that the target disorder is present, and the higher the LR, the greater is this increase.
- LR $<$ 1.0 decreases the probability of the target disorder, and the smaller the LR, the greater is the decrease in probability and the smaller is its final value.

Unlike sensitivity and specificity, LRs can deal with both dichotomous and polychotomous test results. Collapsing multilevel tests into two levels, however, moves LRs closer to 1 and thus results at the extremes will appear less powerful in changing disease probability (see Table 5.9). For a test with only two outcomes, LRs can be calculated directly from sensitivities and specificities. The positive LR is the proportion with disease X who were positive (sensitivity) divided by the proportion without disease X who were positive (1 − specificity). The negative LR is the proportion with disease X who were negative (1 − sensitivity) divided by the proportion without disease X who were negative. However, unlike sensitivity and specificity, computation of LRs does not require dichotomization of test results as depicted in the tables.

Using LRs for Prediction of Disease

We can convert the pretest probability to an odds [probability = odds/(odds + 1)]. We can use Bayes' theorem to convert pretest odds to post-test odds [post-test odds = pretest odds × LR]. Odds can then be converted to probabilities [odds = probability/(1 − probability)] as follows:

- Estimate probability before the test
- Calculate odds before the test

Apply the test

- Calculate odds after the test
- Calculate probability after the test

PPVs and NPVs can also be used to do the same thing, but LRs are easier. LRs have advantages over PPV and NPV because

- They are less likely to change with the prevalence of the disorder
- They can be calculated for several levels of the symptom/sign or test
- They can be used to combine the results of multiple diagnostic tests
- They can be used to calculate the post-test probability for a target disorder.

Table 5.9 Computation of LR from multiple levels of a test result

Serum levels of biochemical test	Disease present	Disease absent	LRs	Serum levels of bio-chemical test	Disease present	Disease absent	LR
≤18	47 (47/85 = 0.553)	2 (2/150 = 0.013)	0.553/0.013 = 42.5	≤45	70 (70/85 = 0.824)	15 (15/150 = 0.1)	0.824/0.1 = 8.24
19–45	23 (23/85 = 0.271)	13 (13/150 = 0.087)	0.271/0.087 = 3.11		Se	1 − Sp	Se/1 − Sp
46–100	7 (7/85 = 0.082)	27 (27/150 = 0.18)	0.082/0.18 = 0.456	>45	15 (15/85 = 0.176)	135 (135/150 = 0.9)	0.176/0.9 = 0.196
>100	8 (8/85 = 0.094)	108 (108/150 = 0.72)	0.094/0.72 = 0.131		1 − Se	Sp	1 − Se/Sp
Totals	85	150		Totals	85	150	

As a rough guide
LRs > 10 or < 0.1 generate large and often conclusive changes from pre- to post-test probability (> 45 % change)
LRs of 5–10 and 0.1–0.2 generate moderate shifts in pre- to post-test probability (about 30–45 % change)
LRs of 2–5 and 0.5–0.2 generate small (but sometimes important) changes in probability (about 15–30 % change)
LRs of 1–2 and 0.5–1 are probably not useful (0–15 % change)

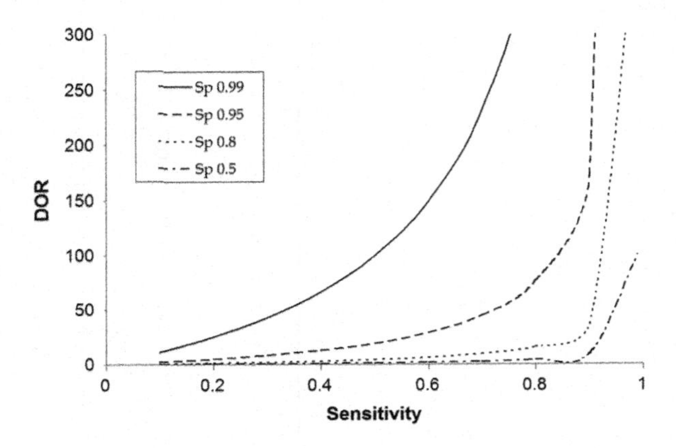

Fig. 5.1 The behavior of the DOR with changing sensitivity and specificity. Each curve represents a fixed specificity denoted in the label and the relationship between DOR and sensitivity is displayed as specificity declines. It can be seen that the DOR increases more rapidly with sensitivity when specificity is high

Diagnostic Odds Ratio

The diagnostic odds ratio (DOR) can be used as a single indicator of test performance. Just like the odds ratio it is the odds of positivity in the abnormal reference group relative to the odds of positivity in the normal reference group.

DOR is given by

$$\mathrm{DOR} = \frac{\mathrm{TP}/\mathrm{FP}}{\mathrm{FN}/\mathrm{TN}} = \frac{\mathrm{LR}(+)}{\mathrm{LR}(-)}$$

The DOR can range between 0 and infinity on a numerical scale and the higher its value, the better the discriminatory performance of the test. Just like the odds ratio, a value of 1 is the null value and implies a non-discriminatory test. Values between 0 and 1 means that the abnormal reference group are more likely to have negative tests than the normal reference group and thus implies a wrong test implementation. The DOR increases steeply when the sensitivity or specificity becomes near perfect (Fig. 5.1). One useful property of the DOR is that it can be derived from binary outcome logistic models (see Chap. 10) that allows the inclusion of additional variables to adjust for heterogeneity. It can also be pooled in meta-analysis but just like the sROC approaches, the true-positive and false-positive rates are modeled simultaneously.

Bibliography

Baicus C, Tanasescu C, Ionescu R (1999) Has this patient a cancer? The assessment of weight loss, anemia and erythrocyte sedimentation rate as diagnostic tests in cancer. A retrospective study based in a secondary care university hospital in Romania. Rom J Intern Med 37:261–267

Doi SA (2011) Understanding evidence in health care: using clinical epidemiology, 1st edn. Palgrave Macmillan, Sydney

Glas AS, Lijmerb JG, Prinsc MH, Bonseld GJ, Bossuyt PM (2003) The diagnostic odds ratio: a single indicator of test performance. J Clin Epidemiol 56:1129–1135

Grimes DA, Schulz KF (2005) Refining clinical diagnosis with likelihood ratios. Lancet 365:1500–1505

Last JM (2001) A dictionary of epidemiology, 4th edn. Oxford University Press, Oxford

Sackett DL, Haynes RB, Guyatt GH, Tugwell P (1991) Clinical epidemiology: a basic science for clinical medicine, 2nd edn. Little, Brown and Company, Boston

Chapter 6
Using and Interpreting Diagnostic Tests with Quantitative Results

The ROC Curve in Diagnostic Accuracy

Suhail A.R. Doi

Abstract Sensitivity and specificity, as defined previously, depend on the cut-off point used to define positive and negative test results. To determine the best cut-off point shift that optimizes sensitivity and specificity, the receiver operating characteristic (ROC) curve is often used. This is a plot of the sensitivity of a test versus its false-positive rate for all possible cut-off points. This chapter outlines its advantages, its use as a means of defining the accuracy of a test, its construction as well as methods for identification of the optimal cut-off point on the ROC curve. Meta-analysis of diagnostic studies is briefly discussed.

Introduction

As we have seen previously, diagnostic test results use a cut-off value based on either a binary or polychotomous scale to define positive and negative test outcomes. On a continuous or ordinal scale, the sensitivity (Se, the probability of a positive test outcome in a diseased individual) and specificity (Sp, the probability of a negative test outcome in a non-diseased individual) can also be computed for specific values. Since the diagnostic test considers the results in two populations, one population with a disease, the other population without the disease, therefore, for every possible cut-off point or criterion value there will be some cases with the disease correctly classified as positive (TP, true-positive fraction), and falsely classified as negative (FN, false-negative fraction). Similarly, cases without the disease can be correctly classified as negative (TN, true-negative fraction) or as positive (FP, false-positive fraction). Compared to the binary outcome, on a continuous scale, the choice of cut-off will affect the degree of false misclassification

S.A.R. Doi (✉)
School of Population Health, University of Queensland, Herston, QLD, Australia

Princess Alexandra Hospital, Brisbane, Australia
e-mail: sardoi@gmx.net

S.A.R. Doi and G.M. Williams (eds.), *Methods of Clinical Epidemiology*, Springer Series on Epidemiology and Public Health,
DOI 10.1007/978-3-642-37131-8_6, © Springer-Verlag Berlin Heidelberg 2013

by the test and thus for the same test, different cut-offs will have different operating characteristics with an inversely related Se and Sp across cut-off values. Thus, for tests that have an ordinal or continuous cut-off, Se and Sp at a single cut-off value do not fully characterise the tests performance which varies across other potential cut-off values. In this situation, a comparison of such diagnostic tests requires independence from the selected cut-off value and this can be addressed via receiver operating characteristic (ROC) analysis. The ROC method has several advantages:

- Testing accuracy across the entire range of cut-offs thereby not requiring a predetermined cut-off point
- Easily examined visual and statistical comparisons across tests
- Independence from outcome prevalence

Example Data

Data from a prospective evaluation of an Australian pertussis toxin (PT) IgG and IgA enzyme immunoassay are used as an example (May et al. 2012). In this study, the accuracy of anti-PT IgG and anti-PT IgA (as normalized optical density) is examined for the diagnosis of pertussis infection with samples taken within 2–8 weeks after onset of symptoms. The gold standard was *Bordetella pertussis* polymerase chain reaction at the first visit.

Basic Principles

The continuous test result is viewed as a multitude of related tests each represented by a single cut-off with each considered to discriminate between two mutually exclusive states, so that we end up with a Se and Sp that are specific to a selected cut-off value. Each cut-off therefore generates a pair of Se and (1 − Sp) and it is these pairs that are then compared via ROC analysis and at each possible cut-off value for the test. Se and (1 − Sp) are essentially equivalent to the true-positive and false-positive proportions, respectively and when we plot Se against (1 − Sp) for various values of the cut-off across the measurement range, this generates the ROC curve. The ROC curve is therefore a plot of the FP probability on the x-axis and the TP probability on the y-axis across several thresholds of a continuous value measured in each subject, with the positive result being assumed for subjects above the threshold. Each point on the curve represents a Se/Sp pair corresponding to a particular cut-off; the latter are also known as the decision threshold or criterion values.

The ROC method is therefore an overall measure (across all possible cut-offs) of diagnostic performance of a test and can be used to compare the diagnostic performance of two or more laboratory or diagnostic tests. The perfect test with perfect discrimination (no overlap in the diseased and healthy distributions) has a

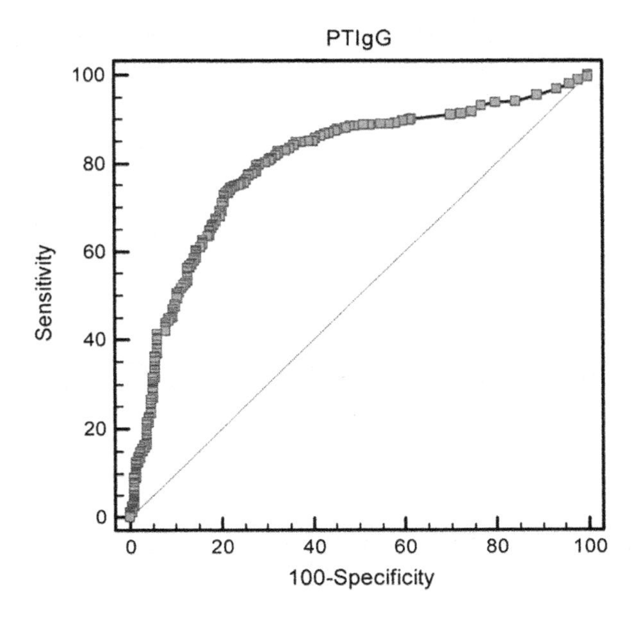

Fig. 6.1 The ROC curve for PT IgG. Each point on the *curve* represents a single cut-off, with sensitivity plotted against 1-specificity (false positive rate). The *central diagonal* is the line of equality

ROC curve that passes through the upper left corner (100 % sensitivity, 100 % specificity). Therefore the closer the ROC curve is to the upper left corner, the higher the overall accuracy of the test. Also, the slope of the tangent line at a cut-point gives the likelihood ratio (LR) for that value of the test. The ROC curve for PT IgG is depicted in Fig. 6.1.

Discrimination Between Diseased and Non-diseased

The area under such an ROC curve is used as a measure of how well a test can distinguish between two diagnostic groups (diseased/normal), independent of any particular cut-off. The closer the curve follows the left-hand border and then the top border of the ROC space, the more area there is under the curve and the more accurate the test. The closer the curve follows the 45° diagonal of the ROC space, the less accurate the test.

The area under the curve (AUC) is therefore a global (i.e. independent of the cut-off value) summary statistic of diagnostic accuracy. The AUC is also known as the c statistic or c index, and can range from 0.5 (random chance or no predictive ability, which would follow the 45° line on the ROC plot) to 1 (perfect discrimination/accuracy; the ROC curve reaches the upper left corner of the graph). The greater the AUC, the more able is the test to capture the trade-off between Se and Sp over a continuous range. According to an arbitrary guideline, one could then use the AUC to classify the accuracy of a diagnostic test (Table 6.1).

Table 6.1 Interpretation of the AUC in terms of accuracy of a test

Accuracy	AUC
Non-informative	AUC = 0.5
Less accurate	0.5 < AUC < 0.7
Moderately accurate	0.7 < AUC < 0.9
Highly accurate	0.9 < AUC < 1
Perfect test	AUC = 1
Results for PT IgG	
Area under the ROC curve (AUC)	0.798
Standard error[a]	0.0177
95 % confidence interval[b]	0.763–0.832
Z statistic	16.836
Significance level P (area = 0.5)	<0.0001

[a]Hanley and McNeil (1982)
[b]AUC ± 1.96 SE

The AUC is interpreted as a probability that a randomly drawn individual from the diseased or abnormal reference sample has a greater test value than a randomly drawn individual from the healthy or normal reference sample. In is clear that if we interpret this as a single member of each group (diseased and non-diseased) taking the test, the probability of a correct answer (the AUC) is not influenced by the prevalence of disease within the sample because each member of the selected pair represents a fixed prevalence at 50 %.

The AUC summarizes the whole of the ROC curve, and therefore all parts of the curve are represented within the AUC. Some parts of the curve can be vertical (lower left part) or horizontal (upper right part) and their contribution to the AUC is less useful because they include cut-off values with increasing Se (without loss of Sp) or increasing Sp (without loss of Se), respectively. Also, we may want to have a fixed Se or Sp for diagnosis (e.g. that Se is at least 80 %), in which case the AUC may not be the best way to compare two tests since the part of the ROC curve below this threshold still contributes to the AUC making this method less optimal for our diagnostic situation.

The 95 % confidence interval is the interval in which the true (population) area under the ROC curve lies with 95 % confidence. The *P* value is the probability that the sample AUC is found when the true (population) AUC is 0.5 (null hypothesis: area = 0.5). If *P* is low (<0.05) then it can be concluded that the AUC is significantly different from 0.5 and therefore there is evidence that the laboratory test does have an ability to distinguish between the two groups. This probability of a correct ranking is the same quantity that is estimated by the non-parametric Wilcoxon statistic and can be used to provide rapid closed-form expressions for the approximate magnitude of the sampling variability, i.e. standard error that one uses to accompany the area under a smoothed ROC curve. Finally, concerning sample size, it has been suggested that meaningful qualitative conclusions can be drawn from ROC experiments performed with a total of about 100 observations. A minimum of 50 cases may be required in each of the two groups, so that one case represents not more than 2 % of the observations.

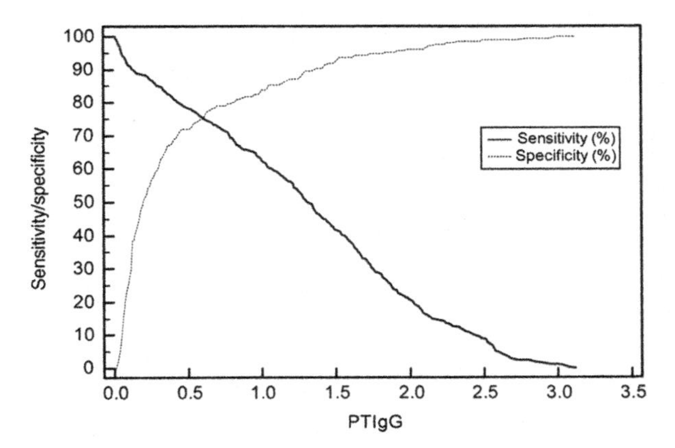

Fig. 6.2 Plot of Se and Sp for different cut-off values of PTIgG (also known as plot vs. criterion values). The criterion was the PCR result

Determining an Optimum Cut-Off Value

If we use the normal distribution to define a value two standard deviations (2SD) above the mean of the normal reference sample, this would result in a cut-off value with a Sp of 97.5 % (since 2SD encompasses 95 % of the population, i.e. 2.5 % on either side of the distribution). This would however not work for skewed or multimodal distributions, and also, it ignores the Se, which is a disadvantage. A better option therefore is to create a table or a plot of Se and Sp for different cut-off values (plot versus criterion values in MedCalc) which can then provide a useful visualization and can also be used to derive two optimal cut-off values: One where good sensitivity is retained and the other where good specificity is retained. This is depicted in the Fig. 6.2 and Table 6.2. Also, it should be kept in mind that the slope of the ROC curve gives us the LR of the test value at that particular cut-off and a table of LR against the cut-off values (see Table 6.2) is an alternative way a cut-off can be selected. Where we choose to place our optimal cut-off will eventually depend on the prevalence of disease in the target population and the consequences of FN versus FP test results (which may differ for every different scenario). For example, a very low prevalence disease with a high cost of false-positive diagnoses may require us to select a cut-off that maximizes Sp. If, on the other hand, for a high prevalence disease where missing a diseased individual has serious consequences, a cut-off value would be selected to maximize Se.

Another alternative is to select the point on the ROC curve closest to the upper left corner of the unit square as this would optimize prevalence-independent summary measures of Se and Sp. The Youden index (Se + Sp − 1) attempts to do this and gives us the optimal or criterion value J corresponding to the maximum of the Youden index; i.e. $J = \max[SE_i + SP_i - 1]$ where SE_i and SP_i are the sensitivity and specificity over all possible threshold values. This value corresponds with the point on the ROC curve farthest from the diagonal line. The MedCalc manual (www.medcalc.org) indicates the following pointers for interpretation of the criterion value:

Table 6.2 Criterion values and coordinates of the ROC curve

Criterion	Sensitivity	95 % CI	Specificity	95 % CI	+LR	95 % CI	–LR	95 % CI
≥0.01	100.00	99.1–100.0	0.00	0.0–1.7	1.00			
>0.05	95.40	92.8–97.2	10.90	7.0–15.9	1.07	0.7–1.6	0.42	0.3–0.7
>0.1	91.30	88.1–93.9	27.49	21.6–34.0	1.26	1.0–1.6	0.32	0.2–0.4
>0.15	89.26	85.8–92.1	41.71	35.0–48.7	1.53	1.3–1.8	0.26	0.2–0.4
>0.2	88.49	84.9–91.5	50.71	43.8–57.6	1.80	1.6–2.1	0.23	0.2–0.3
>0.3	85.17	81.2–88.5	60.66	53.7–67.3	2.17	1.9–2.4	0.24	0.2–0.3
>0.45	79.80	75.5–83.7	71.56	65.0–77.5	2.81	2.5–3.1	0.28	0.2–0.4
>0.64[a]	74.42	69.8–78.7	77.73	71.5–83.2	3.34	3.0–3.7	0.33	0.2–0.4
>0.7	72.89	68.2–77.2	79.15	73.0–84.4	3.50	3.2–3.8	0.34	0.3–0.5
>0.8	69.31	64.5–73.8	80.09	74.1–85.3	3.48	3.2–3.8	0.38	0.3–0.5
>1.35	48.08	43.0–53.2	90.05	85.2–93.7	4.83	4.3–5.4	0.58	0.4–0.9
>1.91	23.53	19.4–28.1	95.26	91.5–97.7	4.96	4.1–6.0	0.80	0.4–1.5
>2.72	2.56	1.2–4.7	99.05	96.6–99.9	2.70	1.5–5.0	0.98	0.2–3.9
>2.8	2.56	1.2–4.7	99.53	97.4–100.0	5.40	2.9–10.0	0.98	0.1–6.9
>2.95	1.28	0.4–3.0	99.53	97.4–100.0	2.70	1.1–6.4	0.99	0.1–7.0
>2.99	1.28	0.4–3.0	100.00	98.3–100.0			0.99	
>3.12	0.00	0.0–0.9	100.00	98.3–100.0			1.00	

[a]Cut-off via the Youden index

- When you select a lower criterion value, then the true-positive fraction and the sensitivity increases. On the other hand, the false-positive fraction also increases, and therefore the true-negative fraction and specificity decrease.
- When you select a higher criterion value, the false-positive fraction decreases with increased specificity but, on the other hand, the true-positive fraction and sensitivity decrease.
- If a test is used for the purpose of screening, then a cut-off value with a higher sensitivity and negative predictive value must be selected. In order to confirm the disease, the cases positive in the screening test can be tested again with a different test. In this second test, a high specificity and positive predictive value are required.

ROC analysis can also be used to evaluate the diagnostic discrimination' of logistic regression models in general as they have binary outcomes. In such an analysis, the power of the model's predicted values to discriminate between positive and negative cases is quantified by the AUC, which is sometimes referred to as the c statistic (or concordance index), and varies from 0.5 (discriminating power not better than chance) to 1.0 (perfect discriminating power). Essentially, we can save the predicted probabilities and use this new variable in ROC curve analysis. The dependent variable used in logistic regression then acts as the classification variable in the ROC curve analysis.

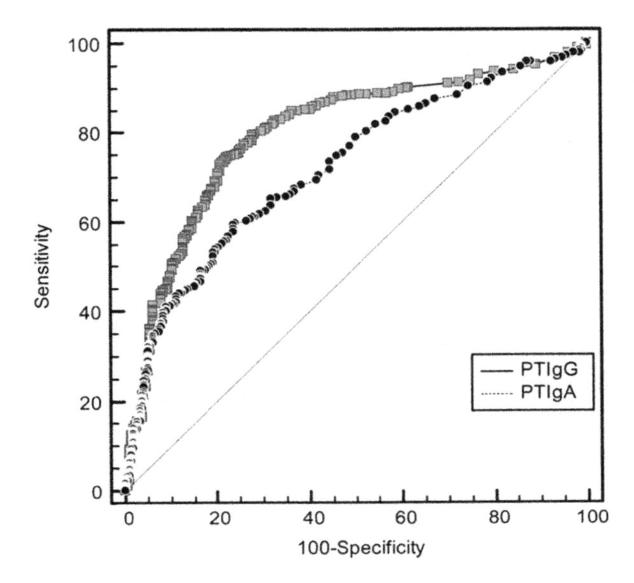

Fig. 6.3 ROC curves comparing plots for PT IgG and PT IgA. Each point on the *curve* represents a single cut-off. The *diagonal* is the line of equality and the higher the plot above this line, the higher its discriminative capacity

Table 6.3 Area under the ROC curves (AUC) for PT IgG and PT IgA

	AUC	SE[a]	95 % CI[b]
PT IgG	0.798	0.0177	0.763–0.832
PT IgA	0.720	0.0206	0.680–0.761

[a]Hanley and McNeil (1982)
[b]AUC ± 1.96SE

Test Comparison

As described above, the AUC represents a summary statistic of the overall diagnostic performance of a test. It makes sense therefore to use the AUC to compare the discriminatory abilities of different tests overall and independent of any specific cut-offs they may have. However, the AUC gives equal weights to the entire ROC curve and it could happen that two tests that differ in terms of optimal sensitivity and specificity have a similar AUC. This may happen when ROC curves cross each other though they may have similar AUC estimates.

The non-parametric area under the plots for the above example (Fig. 6.3) is shown in Tables 6.3 and 6.4. The difference (and 95 % confidence interval) from MedCalc output is shown. The confidence interval for the differences between the tests does not include zero and it can be concluded that there is a statistically significant difference in the AUC estimates and thus the performance of the two tests for pertussis (PT IgG and PT IgA). The better test (PT IgG) is the test with the higher dome and thus greater AUC. It can be seen however that the curves overlap at both ends, suggesting that at these cut-offs, the tests characteristics are identical.

Table 6.4 Pairwise comparison of ROC curves

PT IgG ~ PT IgA	
Difference between areas	0.0775
Standard error[a]	0.0183
95 % confidence interval	0.0417–0.113
Z statistic	4.244
Significance level	$P < 0.0001$

[a]Hanley and McNeil (1982)

Meta-Analysis of Diagnostic Studies

The ROC approach can also be applied to combine multiple estimates of Se and Sp for one test across several primary evaluation studies. The procedure is known as meta-analysis of diagnostic tests. This sort of summary ROC pooling for meta-analysis occurred due to the explosion in the discussion surrounding the implicit threshold across studies of the same radiological investigation, so much so that diagnostic meta-analysis moved from univariate pooling of sensitivity and specificity to summary ROC curves as first defined by Moses et al. (1993) and methods for these have subsequently evolved into hierarchical and bivariate sROC models (see Arends et al. 2008; Reitsma et al. 2005; Rutter and Gatsonis 2001). The basic reasoning was that sensitivity and specificity across studies are negatively correlated and thus study investigators must be using different implicit diagnostic thresholds and thus fit in at different points on an ROC curve. These models were thought to account for the potential presence of a (negative) correlation between sensitivity and specificity within studies and address this explicitly by incorporating this correlation into the analysis. However, Simel and Bossuyt (2009) demonstrate that results from univariate and bivariate methods may be quite similar.

The problem with such an approach is that in reality there may be no implicit diagnostic threshold at play. On the contrary, radiologists might make a diagnosis based on an implicit information size threshold based on the amount of obvious information available to the average radiologist on the image. If images are from very sick persons, they will tend to have a lot more information, thus making it both more likely for a true diagnosis to be made as well as for a false diagnosis to be made. On the other hand, subjects that are not as sick have less information on the image and thus the radiologist will meet the implicit information threshold with difficulty. Thus, while the true-positive rate decreases, so too does the false-positive rate. If we have a set of studies from a varying spectrum of subjects, the Se and Sp are negatively correlated simply on the basis of the varying spectrum of disease – a spectrum effect. There is no change in the implicit diagnostic threshold and chasing such a threshold using sROC models is a questionable pursuit since the goal is ill-defined.

A recent study by Willis (2012) that grouped images by high probability or not according to a trained radiographer and then interpreted by junior doctors revealed exactly this phenomenon. Images with more information content (high probability group) were interpreted with higher Se and lower Sp than the low probability group.

The two groups were said to fit perfectly on the ROC curve and were thus incorrectly interpreted as representing doctors changing their implicit diagnostic threshold rather than a change in the information size from the images making it more difficult for low probability radiographs to meet the information threshold required of the doctors. In such a scenario, the real Se and Sp would actually be the combined Sp and Se based on all radiographs, not high versus low probability on an ROC curve if there are no implicit thresholds.

The same author has previously stated that the Se and Sp may vary between different patient subgroups even when the test threshold remains constant, and this lies at the heart of the concept of the spectrum effect (Willis 2008). The latter effect has not only been mixed up with the concept of an implicit threshold but has also been misleadingly called the spectrum bias (Mulherin and Miller 2002). Such subgroup variation is not a bias and just contributes to heterogeneity across studies; these will lead to estimates of test performance that are not generalizable if the studies are mostly non-representative of their relevant clinical populations.

It has been suggested by Goehring et al. (2004) that in some situations this spectrum effect may lead to a spectrum bias, that is, a distortion of the posterior probability, which can potentially affect the clinical decision. It has been shown that spectrum bias on either a positive or a negative test result can be expressed as the subgroup-specific LR divided by the LR in the overall population of patients (ratio of LR or RLR) and this assessment of spectrum bias has been shown to be independent of the pretest probability. In the usual situation in which sensitivity increases from one patient subgroup to another but the specificity simultaneously decreases, the LRs remain constant and thus while spectrum effects are quite common, spectrum bias is usually not an issue. Nevertheless, despite the absence of bias, sensitivity and specificity on their own may not reflect values that are generalizable to the overall populations that the studies are trying to represent. It has therefore been suggested by Moons et al. (2003) that Se and Sp may have no direct diagnostic meaning because they vary across patient populations and subgroups within populations and thus there is no advantage for researchers in pursuing estimates of a test's Se and Sp rather than post-test probabilities. However, the study by Goehring et al. (2004) clearly demonstrates that the subgroup/population RLR (and thus post-test probability) will not change across subgroups simply due to a spectrum effect because Se and Sp change simultaneously. There is an advantage in pursuing Se and Sp over and above post-test probabilities, and that is to determine, for an average subject of the types represented in the trials, what the expected false-positive and false-negative rates likely to be.

If what we need is the best estimate of Se and Sp across studies that reflects a generalizable population value, then a weighted average of the spectrum of effects across the studies themselves is not necessarily bad. What may lead to spectrum bias, however (as opposed to the spectrum effect), is the methodological rigueur with which the study was conducted and thus a quality assessment is necessary. This is preferable to simply considering the varying effects across studies as random changes because the set of studies cannot be visualized as a random subset from a population of all studies, and therefore the random effects model does not

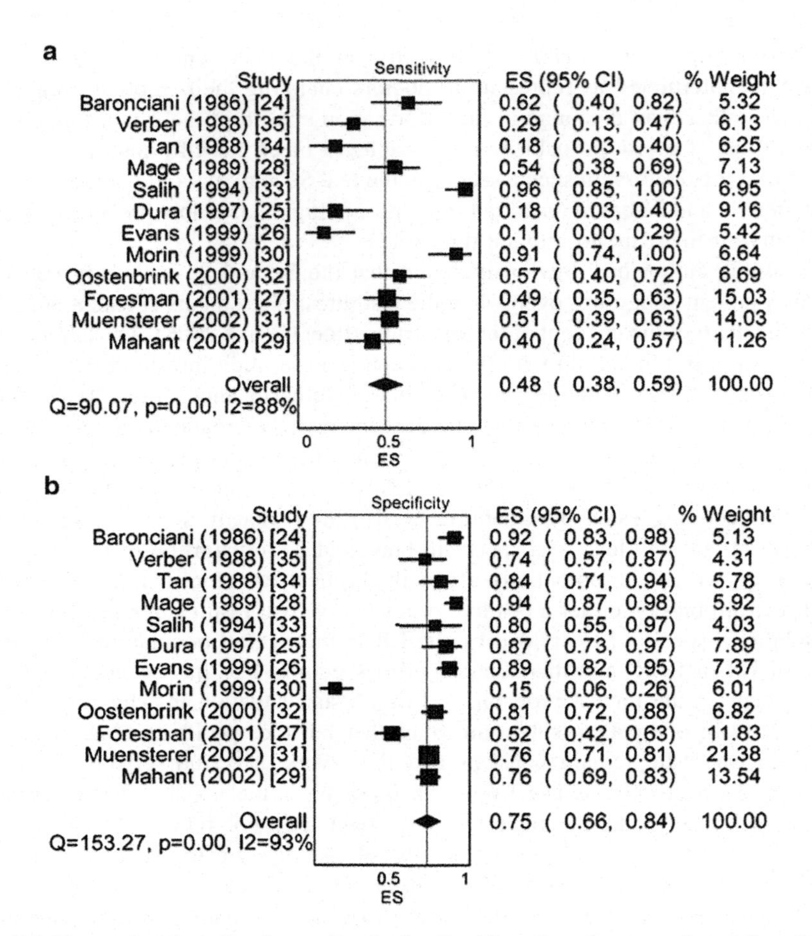

Fig. 6.4 Forest plots depicting the results of univariate bias adjusted meta-analyses of sensitivity and specificity. The box sizes are proportional to the weight given to each study for pooling. The *horizontal lines* are the confidence intervals and the *diamond* depicts the summary sensitivity or specificity. Q = the Cochran Q statistic I2 = the I^2 statistic

apply, at least according to a strict interpretation of randomization in statistical inference. Until recently, there was no simple way of bias adjustment in meta-analysis, but there is currently a quality effects method and software (MetaXL) that implements this (http://www.epigear.com). MetaXL uses a double arcsin square root transformation to stabilize variances of proportions for meta-analysis and provides a method for bias adjustment in addition to the usual inverse variance adjustment.

If we take the example of standard ultrasound data presented by Whiting et al. (2005) for the diagnosis of vesico-ureteral reflux in children, the results of a univariate analysis of Se and Sp are shown in Fig. 6.4. Using the *metandi* procedure in Stata, we can also compute bivariate results and the resulting sROC plot is shown

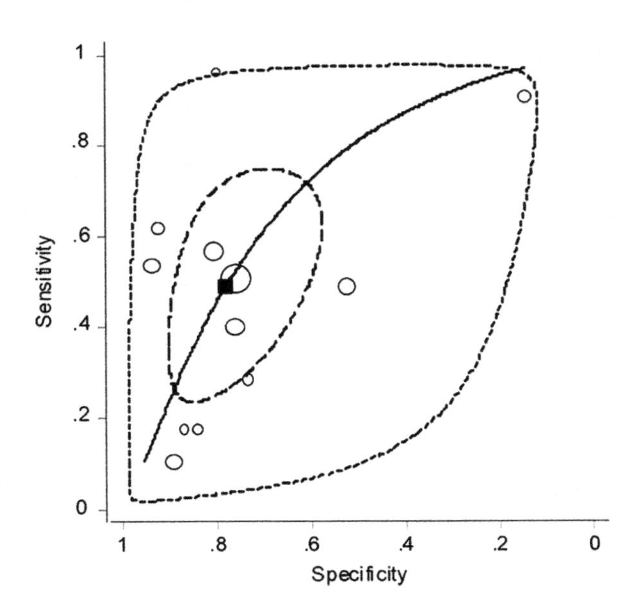

Fig. 6.5 sROC curve generated using the *metandi* procedure in Stata. Study estimates are shown as *circles* sized according to the total number of individuals in each study. Summary (square marker) Se was 49 % (95 % CI 30.6–67.7) and Sp was 78.1 % (95 % CI 64.8–87.3) and the 95 % confidence region for the summary operating point is depicted by the small oval in the centre. The larger oval is the 95 % prediction region (confidence region for a forecast of the true sensitivity and specificity in a future study). The summary curve is from the HSROC model

in Fig. 6.5. The bivariate summary sensitivity was 49 % (95 % CI 30.6–67.7) and the summary specificity was 78.1 % (95 % CI 64.8–87.3). Clearly, these are quite similar to the univariate results in Fig. 6.4 and the added advantage of the univariate plots would be the obvious depiction of the spectrum of effects as well as their correlation with bias, if any.

Bibliography

Arends LR, Hamza TH, van Houwelingen JC, Heijenbrok-Kal MH, Hunink MG, Stijnen T (2008) Bivariate random effects meta-analysis of ROC curves. Med Decis Mak 28:621–638

Gardner IA, Greiner M (2006) Receiver-operating characteristic curves and likelihood ratios: improvements over traditional methods for the evaluation and application of veterinary clinical pathology tests. Vet Clin Pathol 35:8–17

Goehring C, Perrier A, Morabia A (2004) Spectrum bias: a quantitative and graphical analysis of the variability of medical diagnostic test performance. Stat Med 23:125–135

Griner PF, Mayewski RJ, Mushlin AI, Greenland P (1981) Selection and interpretation of diagnostic tests and procedures. Ann Intern Med 94:555–600

Hanley JA, McNeil BJ (1982) The meaning and use of the area under a receiver operating characteristic (ROC) curve. Radiology 143(1):29–36

Harbord RM, Whiting P (2009) Metandi: meta-analysis of diagnostic accuracy using hierarchical logistic regression. Stata J 9:211–229

May ML, Doi SA, King D, Evans J, Robson JM (2012) Prospective evaluation of an Australian pertussis toxin IgG and IgA enzyme immunoassay. Clin Vaccine Immunol 19:190–197

Moons KG, Harrell FE (2003) Sensitivity and specificity should be de-emphasized in diagnostic accuracy studies. Acad Radiol 10:670–672

Moses LE, Shapiro D, Littenberg B (1993) Combining independent studies of a diagnostic test into a summary ROC curve: data-analytic approaches and some additional considerations. Stat Med 12:1293–1316

Mulherin SA, Miller WC (2002) Spectrum bias or spectrum effect? Subgroup variation in diagnostic test evaluation. Ann Intern Med 137:598–602

Reitsma JB, Glas AS, Rutjes AW, Scholten RJ, Bossuyt PM, Zwinderman AH (2005) Bivariate analysis of sensitivity and specificity produces informative summary measures in diagnostic reviews. J Clin Epidemiol 58:982–990

Rutter CM, Gatsonis CA (2001) A hierarchical regression approach to meta-analysis of diagnostic test accuracy evaluations. Stat Med 20:2865–2884

Simel DL, Bossuyt PM (2009) Differences between univariate and bivariate models for summarizing diagnostic accuracy may not be large. J Clin Epidemiol 62:1292–1300

Whiting P, Harbord R, Kleijnen J (2005) No role for quality scores in systematic reviews of diagnostic accuracy studies. BMC Med Res Methodol 5:19

Willis BH (2008) Spectrum bias – why clinicians need to be cautious when applying diagnostic test studies. Fam Pract 25:390–396

Willis BH (2012) Empirical evidence that disease prevalence may affect the performance of diagnostic tests with an implicit threshold: a cross-sectional study. BMJ Open 2:e000746

Chapter 7
Sample Size Considerations for Diagnostic Tests

Application to Sensitivity and Specificity

Rajeev Kumar Malhotra

Abstract Innovations in diagnostic techniques are increasing worldwide. Sensitivity and specificity are used to measure the accuracy of new dichotomous outcome diagnostic tests in the presence of an existing gold standard. The first question to be dealt is what number of subjects is sufficient to attain adequate power in the case of hypothesis testing. This chapter explains sample size issues for estimating the sensitivity and specificity and compares sensitivity and specificity under different goals for two commonly used study designs, case–control and prospective designs in paired and unpaired subjects. The chapter also explains and compares the three methods to control uncertainty under the prospective design.

Introduction

New innovations in diagnostic techniques are increasing due to advancements in technology. Researchers develop newer diagnostic techniques as surrogates for existing gold standards that are difficult to adopt in practice, expensive, not widely available, invasive, risky, and time consuming. For example, the diagnosis of pancreatic carcinoma can be confirmed only by invasive methods such as laparotomy or at autopsy. New diagnostic screening tests for detecting disease should be tested against existing reference diagnostic methods or with an established gold standard. The gold standard may be a more expensive diagnostic method or a combination of tests (combined reference standard) or may be available from clinical follow-up, surgical verification, biopsy, autopsy or a panel of experts. As we have seen in previous chapters, sensitivity and specificity are two indices that measure the accuracy of dichotomous outcome diagnostic tests with respect to an

R.K. Malhotra (✉)
Department of Biostatistics and Medical Informatics, University College of Medical Sciences, New Delhi, India
e-mail: rajeev.kumar.malhotra@gmail.com

S.A.R. Doi and G.M. Williams (eds.), *Methods of Clinical Epidemiology*, Springer Series on Epidemiology and Public Health,
DOI 10.1007/978-3-642-37131-8_7,

existing gold standard. The first question to be dealt with is what is an adequate number of subjects needed for valid estimation of the diagnostic indices (sensitivity or specificity) or to attain adequate power in case of hypothesis testing. This chapter explains the sample size for sensitivity and specificity based on various goals and study designs. Sample size estimation is important not only for adequate power but also for other estimates such as costs and duration of the study. Arbitrary sample size based on convenience fails to achieve adequate power. Small sample sizes produces imprecise estimates and an unduly large sample is a waste of resources especially when the new method is expensive or leads to inconvenience when invasive.

Factors That Influence Sample Size

Four parameters influence the sample size: (1) type I error (probability of rejecting the null hypothesis when it is true), also known as the significance level and denoted by alpha (α) conventionally fixed at 5 %; (2) power, the probability of rejecting the null hypothesis when an alternative hypothesis is true, denoted by $(1 - \beta)$; beta (β) is a type II error and normally assumed to be 0.20 or 0.10 to get 80 % or 90 % power, respectively; (3) expected sensitivity/specificity under the null and alternative hypotheses; this can be obtained from previous studies, a pilot study, or expert opinion; (4) precision or clinically important difference (effect size). The power parameter only influences hypothesis testing. Other factors that may affect the sample size are the study design (case–control, prospective design), paired or unpaired subjects, sampling methods (random or cluster), non-response and cost. This chapter describes sample size estimation for both types of studies (case–control and prospective design) under different goals considering only the four parameters listed above.

Basic Design Considerations

Case–Control Versus Prospective Design

The accuracy of a new diagnostic test can be assessed using a case–control or prospective design. In case–control design, the subjects are required to have true disease status (cases) or non-disease (controls) before conducting the study. In other words, the disease status is known before the new diagnostic test is applied. However, in a prospective design, the status of disease is unknown before conducting the study and subjects are randomly drawn from the population suspected to have the disease. For example, to apply a new diagnostic test (magnetic resonance imaging (MRI)) for the detection of prostate cancer, the researcher should apply both the gold standard, such as histopathology, or new diagnostic MRI

Table 7.1 Comparison of case–control and prospective designs in diagnostic studies

	Case–control design	Prospective design
Sample size	Small sample size is required because disease status is already known. Thus, the researcher has control over the sample size	Large sample size is required because sample size depends on the prevalence of disease, especially when the prevalence of disease is low
Estimated indicators	Only sensitivity and specificity are valid indicators	Apart from sensitivity and specificity, positive and negative predictivity can also be calculated
Effect of severity of disease on the study	The severity of disease can affect the sensitivity and specificity. This bias is called spectrum bias. Nevertheless, this bias can be improved by using a nested case–control study, e.g. the stage of cancer may affect the new diagnostic accuracy. The new diagnostic test may be more sensitive in advanced stages of cancer than in initial stages of cancer. Accuracy in such situations needs to be checked for each cancer stage separately	Since the cohort is a random sample from the population, it is free from spectrum bias
Cost of diagnostic test	Less costly because true status of disease is already known	More costly because both the gold standard and the new diagnosis test are applied to each randomly selected subject

to a randomly selected group of patients suspected to have prostate cancer. In a prospective design, the gold standard is determined subsequent to the study. A comparison between case–control and prospective designs is shown in Table 7.1.

Paired and Unpaired Designs

Another important issue arises when the goal is to compare two new diagnostic test indices in the presence of the existing gold standard. In paired designs, both tests are applied to each subject, whereas, in the case of unpaired designs, different subjects are assigned to each test. Generally, a paired design is preferable to an unpaired design because it reduces the sample size and eliminates subject-to-subject variation. A paired study is usually statistically more powerful than an unpaired study. However, the results are biased if the researcher already knows the result of one test. Thus, to avoid such bias, tests should be performed by two independent experts with comparable experience.

Unpaired designs are useful when one or both of the tests are either invasive or inconvenient. In such situations, a parallel design is ideal. In a parallel design,

Table 7.2 Common notation used in sample size formulas and their explanation

Notation	Explanation
n_D	Number of subjects with the disease or condition required for the study. i.e. cases
n_{ND}	Number of subject with non-disease or without the condition required for the study, i.e. controls
n	Total sample size for the study
S_{N0}	Pre-specified value of sensitivity under the null hypothesis
S_N	Expected sensitivity of a diagnostic test
S_{N1}	Sensitivity of first diagnostic test
S_{N2}	Sensitivity of second diagnostic test
S_{P0}	Pre-specified value of the specificity under the null hypothesis
S_P	Expected specificity of a diagnostic test
S_{P1}	Specificity of the first diagnostic test
S_{P2}	Specificity of the second diagnostic test
$Z_{1-\alpha/2}$	Percentile of standard normal distribution at the α level of significance (two-tailed)
$Z_{1-\beta}$	Percentile of standard normal distribution at β, where β is the type II error and $1-\beta$ is the power
ω	Proportion of disagreement in case of a paired design study
Prev	Anticipated prevalence of disease in population suspected to have the disease
L	Precision (half the length of the confidence interval)

subjects are randomized to the diagnostic test they receive and care must be taken with the factors that can influence the results of test. These factors could be the severity of disease (cancer stage), experience of the tester, age, sex, race, region etc.

Notation

In this chapter, the sample size formulas are presented for two-tailed hypotheses (two-tailed is a common selection process; it is used to test whether sensitivities differ irrespective of which proportion is larger); for a one-tailed hypothesis, the researcher should use $Z_{1-\alpha}$ instead of $Z_{1-\alpha/2}$. Assuming the large sample theory, a normal approximation to binomial distribution is used in all situations except the exact case. Table 7.2 presents the common notation used in the sample size formulas throughout this chapter.

Sample Size for a Case–Control Design

Sample Size for Estimating Diagnostic Indices (Sensitivity or Specificity or Both) with a Given Precision

Sample size is necessary to estimate the expected sensitivity or specificity of a new diagnostic test in the presence of a gold standard and is calculated using the confidence interval (CI) approach:

$$S_{\mathrm{N}} \pm Z_{1-\alpha/2}\sqrt{\frac{S_{\mathrm{N}} \times (1 - S_{\mathrm{N}})}{n_{\mathrm{D}}}}$$

This CI shows that the true value of the sensitivity can be as low as

$$S_{\mathrm{N}} - Z_{1-\alpha/2}\sqrt{\frac{S_{\mathrm{N}} \times (1 - S_{\mathrm{N}})}{n_{\mathrm{D}}}}$$

and as high as

$$S_{\mathrm{N}} + Z_{1-\alpha/2}\sqrt{\frac{S_{\mathrm{N}} \times (1 - S_{\mathrm{N}})}{n_{\mathrm{D}}}}$$

The sample size formula based on normal approximation to binomial in a case–control design is

$$n_{\mathrm{D}} = \frac{Z_{1-\alpha/2}^2 \times S_{\mathrm{N}} \times (1 - S_{\mathrm{N}})}{L^2} \tag{7.1}$$

The experimenter specifies the desired width of the CI with an α confidence level normally taken as 95 %. Continuity correction applies when continuous distribution is used as an approximation to a discrete distribution such as normal approximation to binomial. The CI with $100(1 - \alpha)\%$ with continuity correction is

$$S_{\mathrm{N}} \pm \left[Z_{1-\alpha/2}\sqrt{\frac{S_{\mathrm{N}} \times (1 - S_{\mathrm{N}})}{n_{\mathrm{D}}}} + \frac{1}{2n_{\mathrm{D}}}\right]$$

To calculate the sample size with continuity correction, sample size software such as NCSS PASS version-11 or MS Excel 2007 can be used. Normal approximation formula may not be adequate when the sensitivity or specificity is more than 0.95. In this situation, an exact binomial method described later in this chapter should be applied. Table 7.3 in the Appendix provides the sample size for different values of sensitivity or specificity using exact, normal approximation, and with

Table 7.3 Number of subjects with disease (or without disease) for estimation of the expected sensitivity (or specificity) ranging from 0.50 to 0.99

Sensitivity/ specificity	Sample size based on exact (Clopper–Pearson)									
	±0.01	±0.02	±0.03	±0.04	±0.05	±0.06	±0.07	±0.08	±0.09	±0.10
0.50	9,701	2,449	1,098	623	402	281	208	160	127	104
0.55	9,605	2,425	1,088	617	398	279	206	159	126	103
0.60	9,317	2,353	1,056	599	397	271	200	154	123	100
0.65	8,837	2,233	1,002	569	367	257	191	147	117	95
0.70	8,165	2,065	928	527	341	239	177	137	109	89
0.75	7,301	1,849	832	473	306	215	159	123	98	80
0.80	6,245	1,585	715	407	264	186	138	107	85	70
0.85	4,997	1,273	576	330	215	151	113	88	70	58
0.90	3,557	914	417	241	158	112	84	66	53	44
0.95	1,927	508	238	140	94					
0.96	1,578	422	200	119						
0.97	1,223	334	161							
0.98	861	245								
0.99	497									

Sensitivity/ specificity	Sample size based on normal approximation to binomial									
	±0.01	±0.02	±0.03	±0.04	±0.05	±0.06	±0.07	±0.08	±0.09	±0.10
0.50	9,604	2,401	1,068	601	385	267	196	151	119	97
0.55	9,508	2,377	1,057	595	381	265	195	149	118	96
0.60	9,220	2,305	1,025	577	369	257	189	145	114	93
0.65	8,740	2,185	972	547	350	243	179	137	108	88
0.70	8,068	2,017	897	505	323	225	165	127	100	81
0.75	7,203	1,801	801	451	289	201	147	113	89	73
0.80	6,147	1,537	683	385	246	171	126	97	76	62
0.85	4,898	1,225	545	307	196	137	100	77	61	49
0.90	3,458	865	385	217	139	97	71	55	43	35
0.95	1,825	457	203	115	73					
0.96	1,476	369	164	93						
0.97	1,118	280	125							
0.98	753	189								
0.99	381									

Sensitivity/ specificity	Sample size based on normal approximation to binomial (with continuity correction)									
	±0.01	±0.02	±0.03	±0.04	±0.05	±0.06	±0.07	±0.08	±0.09	±0.10
0.50	9,704	2,451	1,101	625	404	284	211	163	130	106
0.55	9,608	2,427	1,090	619	401	281	209	161	129	105
0.60	9,320	2,355	1,058	601	389	273	203	157	125	102
0.65	8,840	2,235	1,005	571	370	260	193	149	119	98
0.70	8,167	2,067	930	529	343	241	179	139	111	91
0.75	7,303	1,851	834	475	308	217	161	125	100	82
0.80	6,246	1,587	716	409	266	188	140	109	87	72
0.85	4,998	1,274	578	331	216	153	114	89	72	59
0.90	3,557	914	417	241	158	113	85	66	54	45
0.95	1,924	505	235	138	92					
0.96	1,574	418	196	116						

(continued)

Table 7.3 (continued)

Sensitivity/ specificity	Sample size based on normal approximation to binomial (with continuity correction)									
	±0.01	±0.02	±0.03	±0.04	±0.05	±0.06	±0.07	±0.08	±0.09	±0.10
0.97	1,216	328	156							
0.98	850	236								
0.99	476									

NCSS PASS software was used to calculate the sample size

continuity correction. The sample size obtained using continuity correction was close to the sample size calculated the using exact method. When the goal is to estimate the specificity, just replace S_N by S_P and n_D by n_{ND}.

If sensitivity is important for the study, determine the number of disease positive subjects for the expected sensitivity and select the same number of disease negative subjects. If the specificity is important, determine the number of disease negative subjects for the expected specificity and select the same number of disease positive subjects. However, when sensitivity and specificity are equally important and have different expected values, determine the number of subjects based on both indices separately; the study sample size is the sum of the sample sizes obtained by these indices.

In some situations, it is not possible to estimate the sensitivity or specificity in advance. Then, the safest option for determining the sample size is to assume the expected sensitivity or specificity is 0.50 with 0.05 or 0.10 absolute precision either side because that gives the maximum sample size. This works well when the anticipated sensitivity or specificity lies between 10 % and 90 %. However, it should not be used when the expected sensitivity or specificity is >0.95; the exact binomial method is preferred to determine the sample size in this case.

Example 1 Estimate an adequate sample size to determine diagnostic accuracy of computed tomography (CT) for prostate cancer in males greater than 60 years of age. Previous literature shows that CT has 85 % sensitivity and 90 % specificity. Absolute precision either side of the 95 % confidence level is 5 %. Both sensitivity and specificity are equally important. The gold standard to diagnosis prostate cancer is histopathology.

Solution Using Eq. 7.1 and $Z_{1-\alpha/2} = 1.96$, $S_N = 0.85$, $L = 0.05$, the number of subjects with disease confirmed required is

$$n_D = \frac{Z_{1-\alpha/2}^2 \times S_N(1 - S_N)}{L^2} = \frac{1.96^2 \times 0.85\ (1 - 0.85)}{0.05^2} = 196$$

The number of non-disease subjects required is

$$n_{ND} = \frac{Z_{1-\alpha/2}^2 S_P\ (1 - S_P)}{L^2} = \frac{1.96^2 \times 0.90\ (1 - 0.90)}{0.05^2} = 139$$

Thus, 335 subjects would be needed to estimate the sensitivity/specificity with 95 % confidence level and estimates within ±0.05 of the true value. With continuity correction, there is a need for 374 subjects (216 with disease and 158 without disease subjects) for the study.

Sample Size for Detecting a Minimum Clinically Important Difference in the Sensitivity/Specificity of a Single Test

Sensitivity S_N follows binomial distribution (n, S_{N0}) under the null hypothesis and binomial distribution (n, S_{N1}) under an alternative hypothesis. Normal approximation is assumed for this binomial distribution (assuming large n). The sample size formula for sensitivity under the hypotheses H_0: $S_N = S_{N0}$ and H_A: $S_N \neq S_{N0}$ (two-tailed) is described by Arkin and Wachtel (1990).

$$n_D = \frac{\left(Z_{1-\alpha/2}\sqrt{S_{N0}(1-S_{N0})} + Z_{1-\beta}\sqrt{S_{N1}(1-S_{N1})}\right)^2}{\left(S_{N1}-S_{N0}\right)^2} \tag{7.2}$$

Example 2 A study plans to test the sensitivity of breast ultrasonography to detect breast cancer considering histopathology as the gold standard. The previous literature showed that the sensitivity is 0.80 and specificity is 0.85. The investigator wants to know the total sample size to detect a difference of 0.05 in sensitivity and specificity rejecting the null hypothesis with probability 0.9 (90 % power) and keeping the level of significance at 0.05 (95 % CI).

Solution H_0: $S_N = 0.80$ and H_a: $S_N \neq 0.80$. Using Eq. 7.2 and $Z_{1-\alpha/2} = 1.96$, $S_{N0} = 0.80$, $|S_{N1} - S_{N0}| = 0.05$, the number of subjects with disease confirmed is

$$n_D = \frac{\left[1.96 \times \sqrt{0.8 \times 0.2} + 1.28\sqrt{0.85 \times 0.15}\right]^2}{0.05^2} = 616$$

Using Eq. 7.2 and changing S_{N0} to S_{P0} and n_D to n_{ND} with $|S_{P1} - S_{P0}| = 0.05$, the sample size for the specificity is

$$n_{ND} = \frac{\left[1.96 \times \sqrt{0.85 \times 0.15} + 1.28\sqrt{0.90 \times 0.10}\right]^2}{0.05^2} = 470$$

Thus, 616 diseased subjects would be needed to detect a difference 0.05 in sensitivity, with 90 % power and 95 % confidence level. Similarly, 470 non-diseased subjects would be required to detect a difference of 0.05 in specificity from the expected specificity of 0.85 with 90 % power and 95 % confidence level. The study sample (n) would be 1,086.

Sometimes, investigators fix the lower CI limit of a new diagnostic test. The null hypothesis H_0: $S_N = S_{NL}$ against the alternative hypothesis H_a: $S_N > S_{NL}$. S_{NL} is the lower limit for sensitivity with $1 - \alpha$ confidence. The other situation, $S_N < S_{NL}$, may not arise because our concern is that the new diagnostic test does not perform better than the other method. The sample size formula suggested by Machin et al. (1997) is

$$n_{\mathrm{D}} = \frac{\left[Z_{1-\beta}\sqrt{S_{\mathrm{N}}(1-S_{\mathrm{N}})} + Z_{1-\alpha}\sqrt{(S_{\mathrm{N}}-\delta)(1-S_{\mathrm{N}}+\delta)}\,\right]^2}{\delta^2}$$

The expected sensitivity of the diagnostic test is S_{N} and we wish for $1-\alpha$ lower confidence for S_{N} to be more than $S_{\mathrm{N}}-\delta$ with power $1-\beta$. δ is the difference between S_N and S_{NL}. Flahault et al. (2005) provide a sample size table to find the number of subjects using the exact method.

Example 3 The expected sensitivity of a diagnostic test is 0.90. Consider the lower limit for expected sensitivity should not fall below 0.80 with 80 % power and 95 % level of confidence. Calculate the number of disease positive subjects required for the study.

Solution The expected sensitivity $S_{\mathrm{N}} = 0.90$, $1-\beta = 0.90$ and $Z_{1-\beta} = 1.28$, $1-\alpha = 0.95$ and $Z_{1-\alpha} = 1.645$, $\delta = 0.10$

$$n_{\mathrm{D}} = \frac{\left[1.28 \times \sqrt{0.90 \times 0.10} + 1.645 \times \sqrt{0.80 \times 0.20}\,\right]^2}{0.10^2} = 109$$

Thus, 109 confirmed subjects are needed so that the sensitivity does not fall below 0.80 with 90 % power and 95 % confidence level.

Sample Size Formula to Compare Two Diagnostic Tests in the Presence of a Gold Standard

Unpaired Design

The sample size determination for comparison of two diagnostic tests indices is different in unpaired and paired studies. The formulas for the specificity are similar, therefore, only the formula and method for the sensitivity is presented in this chapter. S_{N1} and S_{N2} are the sensitivities of two diagnostic tests. The null and alternative hypotheses are $\mathrm{H_0}$: $S_{\mathrm{N1}} = S_{\mathrm{N2}}$ versus $\mathrm{H_A}$: $S_{\mathrm{N1}} \neq S_{\mathrm{N2}}$ for a large sample and normal approximation. The formula recommended for determination of the sample size is

$$n_{\mathrm{D}} = \frac{\left[Z_{1-\alpha/2}\sqrt{2\bar{S}_{\mathrm{N}}(1-\bar{S}_{\mathrm{N}})} + Z_{1-\beta}\sqrt{S_{\mathrm{N1}}(1-S_{\mathrm{N1}}) + S_{\mathrm{N2}}(1-S_{\mathrm{N2}})}\,\right]^2}{(S_{\mathrm{N1}}-S_{\mathrm{N2}})^2} \tag{7.3}$$

where the mean of two expected sensitivities $\bar{S}_{\mathrm{N}} = (S_{\mathrm{N1}} + S_{\mathrm{N2}})/2$.

Casagrande and Pike (1978) recommended a continuity correction in the above formula

$$n_D(\text{Continuity correction}) = \frac{n_D \times \left[1 + \{1 + 4/(|(S_{N1} - S_{N2})| \times n_D)\}^{1/2}\right]^2}{4}$$

A similar formula is used for the specificity; S_{N1} and S_{N2} are replaced by S_{P1} and S_{P2}, respectively, and n_D is replaced by n_{ND}.

When a diagnostic test has already been performed or the investigator has fixed the number of subjects for the diagnostic test in advance, the sample size for another diagnostic test can be determined by first applying Eq. 7.3 to determine sample size n_D and then applying the formula suggested by Cohen (1977) to get the sample size for the other diagnostic test:

$$\text{sample size for second test} = \frac{n_D \times m}{2 \times m - n_D}$$

where m is number of subjects already fixed by the researcher for the first diagnostic test. If the specificity is important, change the notation accordingly.

Sample Size for a Paired Design

The sample size method for a paired design proposed by Conner (1987) for McNemar's test comparing the two proportions is applied to determine the sample size for comparing sensitivities or specificities. The null hypothesis is H_0: $S_{N1} = S_{N2}$ and the alternative hypothesis H_a: $S_{N1} \neq S_{N2}$. The sample size formula as proposed by Conner (1987) is

$$n_D = \frac{\left[Z_{1-\alpha/2} \times \omega^{1/2} + Z_{1-\beta} \times \left(\omega - (S_{N1} - S_{N2})^2\right)^{1/2}\right]^2}{(S_{N1} - S_{N2})^2} \tag{7.4}$$

where n_D is number of subjects with confirmed disease, S_{N1} and S_{N2} are the expected sensitivities and $S_{N1} - S_{N2}$ is the absolute difference under the alternative hypothesis. ω is the proportion of disagreement. The proportion of disagreement can be estimated from the previous literature, but when the investigator has no idea about this, Conner (1987) proposed an approximation formula to obtain the minimum and maximum proportion of disagreement:

$$\omega = |S_{N1} - S_{N2}| \quad \text{(minimum proportion of disagreement)}$$

$$\omega = S_{N1}(1 - S_{N2}) + S_{N2}(1 - S_{N1}) \quad \text{(maximum proportion of disagreement)}$$

The value of ω close to or equal to the maximum proportion of disagreement produces a large sample of estimates and one can use the mid-point level that lies between the minimum and maximum proportion of disagreement:

$$\omega = \frac{(S_{N1} - S_{N2}) + ((S_{N1}(1 - S_{N1}) + S_{N2}(1 - S_{N2}))}{2}$$

A similar formula can be applied for the specificity in a paired study design:

$$n_{ND} = \frac{\left[Z_{1-\alpha/2} \times \omega^{1/2} + Z_{1-\beta} \times \left(\omega - (S_{P1} - S_{P2})^2\right)^{1/2}\right]^2}{(S_{P1} - S_{P2})^2} \tag{7.5}$$

where n_{ND} is the number of subjects with confirmed disease, S_{P1} and S_{P2} are the expected sensitivities and $S_{P1} - S_{P2}$ is the absolute difference under the alternative hypothesis.

$$\omega = |S_{P1} - S_{P2}| \quad \text{(minimum proportion of disagreement)}$$

$$\omega = S_{P1}(1 - S_{P2}) + S_{P2}(1 - S_{P1}) \quad \text{(maximum proportion of disagreement)}$$

Both sensitivity and specificity formulas assumes that the status of disease and non-disease is already known. In a case–control study where both sensitivity and specificity are important criteria, the study sample size for a paired designed is the sum of diseased subjects and non-diseased subjects $n_D + n_{ND}$.

Example 4 An investigator wants to compare the sensitivity of two diagnostic tests: total adenosine deaminase activity (ADA) and lymphocyte test with cut-off values of >39 U/L and >89 %, respectively, for diagnosing tuberculosis. The expected sensitivity of the ADA and lymphocyte tests is 80 % and 10 % absolute difference in the expected sensitivity of lymphocytes test is considered significant with 80 % power and a 5 % level of significance. The gold standard is a combination of existing methods such as bacteriology and histopathology findings. Compare the sample size obtained from both unpaired and paired designs with minimum, maximum, and mid-point proportion of disagreement, assuming the investigator does not have any idea about the proportion of disagreement.

Solution The absolute difference between the two sensitivities is 10 %. Then H_0: $S_{N1} = S_{N2} = 0.80$ and under the alternative hypothesis H_a: $S_{N1} = 0.75$ and $S_{N2} = 0.85$. The value of the minimum proportion of disagreement is $\omega = 0.10$ and the maximum disagreement is

$$\omega = 0.85 \times (1 - 0.75) + 0.75 \times (1 - 0.85) = 0.325$$

Assuming the lowest proportion of agreement, the estimate for the sample size for sensitivity using Eq. 7.4 for $\omega = 0.10$, $S_{N1} - S_{N2} = 0.10$, $Z_{1-\alpha/2} = 1.96$ at $\alpha = 0.05$, and $Z_{1-\beta} = 0.845$ for $\beta = 0.20$ is

$$n_D = \frac{\left[Z_{1-\alpha/2} \times \omega^{1/2} + Z_{1-\beta} \times \left(\omega - (S_{N1} - S_{N2})^2\right)^{1/2}\right]^2}{(S_{N1} - S_{N2})^2}$$

$$n_D = \frac{\left[1.96 \times (0.1)^{1/2} + 0.845 \times \left(0.1 - (0.1)^2\right)^{1/2}\right]^2}{0.1^2} = 76.27 \cong 77$$

Round to the nearest integer to ensure 80 % power, 77 subjects would be needed at the lowest level of proportion of agreement between the two diagnostic methods.

The sample size requirement for a paired study, assuming the maximum disagreement $\omega = 0.325$ with the same power, confidence level and absolute difference, is

$$n_D = \frac{\left[1.96 \times (0.325)^{1/2} + 0.845 \times \left(0.325 - (0.1)^2\right)^{1/2}\right]^2}{0.1^2} = 253.33 \cong 254$$

Thus, 254 subjects would be required to detect a difference of 0.10 between the two diagnostic tests for sensitivity with 80 % power and 5 % type I error. The sample size based on the mid-point level is 165 subjects. The mid-point level sample size is a compromise between the maximum and minimum sample size.

The sample size requirement for an unpaired study design using Eq. 7.3 is

$$n_D = \frac{\left[1.96 \times \sqrt{2 \times 0.8 \times (1-0.2)} + 0.845\sqrt{0.85 \times (1-0.85) + 0.75 \times (1-0.75)}\right]^2}{0.1^2}$$

$$n_D = \frac{[1.1087 + 0.4743]^2}{0.01} = 251$$

Thus, for an unpaired study design, a sample of 502 (251 in each group) would be required per diagnostic test with similar power, level of significance, and difference. After applying continuity correction, a sample size of 271 per group is required to detect a difference of 0.10 between the sensitivities with 80 % power and 95 % confidence level.

Example 5 Suppose an ADA test has already been performed in 300 patients and the investigator needs an adequate sample for a lymphocytes test to compare the diagnostic sensitivity of ADA with lymphocytes as discussed in Example 4.

Solution Using the formula for sample size for a second test,

$$\text{sample size for second test} = \frac{n_D \times m}{2 \times m - n_D} = \frac{251 \times 300}{2 \times 300 - 251} = \frac{75600}{349} = 216$$

Thus, a sample size of 300 and 216 confirmed subjects for test 1 and test 2, respectively, is required to detect a difference of 0.1 between the sensitivities with 80 % power and 95 % confidence level.

Sample size for comparison of two sensitivities for a paired design may require at least 77 subjects (for minimum proportion of disagreement) and as many as 254 subjects (for maximum proportion of disagreement). Thus, sample size depends on the proportion of disagreement between two diagnostic tests. However, in an unpaired study design, the sample size for a similar problem would be 502 (251 for ADA and 251 for lymphocytes). Thus, the paired design requires 248–425 less disease subjects than the unpaired design.

In the absence of any exact proportion of disagreement, it is advisable to use a sample close to 254 to ensure adequate power of the study. When the diagnostic tests have higher cost, it is beneficial to perform a pilot study to estimate the proportion of disagreement and get the adequate sample size or use the mid-point level approach.

Sample Size Estimation in a Prospective Design

The sample size formula for both case–control and prospective designs are similar except for the role of prevalence. In the case–control study the researcher has control over the disease positive and disease negative cases in advance, whereas in a prospective design, the researcher has no prior control over the disease positive and disease negative subjects. Thus, in a prospective design, prevalence plays a vital role in calculating the sample size. In such situations, first estimate the sample size using the case–control formula and then consider the prevalence effect. When sensitivity and specificity are equally important for the study, determine the sample size separately for both and use the larger sample size. Uncertainty of prevalence in the prospective design can be dealt with using three approaches: the naive approach, the normal distribution approach, and the exact binomial approach. A brief description of these approaches is as follows.

Naive Approach

The naive approach was proposed by Buderer (1996) and the sample size of the study (n) is determined by dividing (n_D) by the expected prevalence of disease for sensitivity. In the case of specificity, divide n_{ND} by (1 − prevalence). This approach is less conservative because it does not incorporate the variation associated with true prevalence and provide the average disease positive number of subjects. In other words, there is approximately 0.50 probability that the number of diseased subjects will be adequate for the study, that is $P((n \times \text{prev}) \geq n_D/n, \text{prev}) = 0.50$ (approximate). Malhotra and Indrayan (2010) devised a nomogram for estimation of sensitivity and specificity of medical tests for various prevalences and for four absolute precisions ±0.03, ±0.05, ±0.07, ±0.10.

Normal Distribution

The true prevalence of disease may be smaller than anticipated and consequently, the study may have fewer numbers of diseases positive subjects compared with the required number of disease positive subjects (i.e. $n \times \text{prev} < n_D$) and affect the power of the study. To get the sufficient number of diseased subjects, the researcher has to find the minimum total sample size n such that the probability $n \times \text{prev} > n_D$ is not less than φ. To control the uncertainty, the following formula, which is an approximation of the binomial distribution (n, prev) to the standard normal distribution was proposed by Schatzkin et al. (1987)

$$Z_\varphi = \frac{n_D - n \times \text{prev}}{\sqrt{\text{prev} \times (1 - \text{prev}) \times \text{n}}}$$

where n is the total sample size after adjusting for the uncertainty in the prevalence of the study, prev is the prevalence of disease in suspected subjects, φ is the preselected probability to ensure the number of diseased subjects and Z_φ is the standard normal variate corresponding to the upper tail of probability φ. The above equation can be easily solved using the MS Excel 2007 Goal Seek command described in detail in the Appendix.

Exact Binomial

To calculate the total sample size so that the sample ensures that, with probability φ, n_D is greater than the target, solves this equation in n to obtain n_D.

$$P((n \times \text{prev}) \geq n_D/n,\ \text{prev}) = \varphi$$

This equation can also be solved using the MS Excel 2007 Goal Seek command, but sometimes Goal Seek fail to get the exact solution. The procedure to solve the exact binomial is given in detail in the Appendix.

Example 6 Using the same parameters as in Example 1 and taking an anticipated prevalence of 30 % for suspected subjects of prostate cancer, what is the total sample size required using the naïve, normal and exact binomial approaches?

Solution The number of disease subjects required in Example 1 is 196 and the anticipated prevalence is 0.3.

Total sample (n) for sensitivity: $\frac{n_D}{\text{prev}} = \frac{196}{0.3} \cong 654$

Total sample (n) for specificity: $\frac{n_{ND}}{1 - \text{prev}} = \frac{139}{0.7} = 198.57 \cong 199$

The sample size is larger for sensitivity than specificity, thus the total sample size for the study is 654 subjects which is adequate for a prospective design.

Normal Approximation Method

To get a total sample size such that the number of diseased subjects (n_D) is greater than 196 with probability ($\varphi = 0.95$) and anticipated diseased prevalence $= 0.30$. Since $\varphi = 0.95$, then $Z_\varphi = -1.645$

$$-1.645 = \frac{196 - n \times 0.3}{\sqrt{0.3 \times (1 - 0.7) \times n}}$$

The total sample size using the normal approximation approach is 721 (approximate). In the case of specificity, the total study subjects is calculated such that the number of non-disease subjects is greater than 139 with preselected probability of 0.95 and anticipated prevalence of 0.7.

$$-1.645 = \frac{139 - n \times 0.7}{\sqrt{0.7 \times (1 - 0.3) \times n}}$$

When specificity is important, 214 subjects are needed. However, if both indices are important, then the higher of the two sample sizes is the total sample size (n).

Exact Binomial Approach

Under similar conditions as described for the normal approximation method, $\varphi = 0.95$ and Bin(n,0.3)

$$P((n \times 0.3 \geq 196/n,\ 0.3) = 0.95$$

For the total sample size using the exact binomial approach, 720 subjects is sufficient with 0.95 probability that the diseased subjects in the study would be 196 or more. Table 7.4 in the Appendix provides the sample sizes for different ranges of sensitivities and various prevalence rates.

Sample Size to Test the One-Sided Equivalence (Non-inferiority) of Two Sensitivities (Paired Design)

The equivalence of sensitivities between a new diagnostic test and the gold standard or between two diagnostic tests sensitivities can be tested. This situation commonly arises if the standard test is expensive, invasive, or inconvenient. The physician or

Table 7.4 Study sample size for estimating the sensitivity in prospective design using the naive method, normal approximation, and exact binomial

Expected sensitivity	Prevalence	Method 1(naïve)			Method 2(normal)			Method 3(exact)		
		±0.05	±0.07	±0.10	±0.05	±0.07	±0.10	±0.05	±0.07	±0.10
0.60	0.05	7,380	3,780	1,860	8,023	4,248	2,196	8,007	4,231	2,180
	0.10	3,690	1,890	930	4,003	2,118	1,094	3,995	2,110	1,087
	0.15	2,460	1,260	620	2,662	1,407	726	2,657	1,403	721
	0.20	1,845	945	465	1,992	1,052	542	1,990	1,049	538
	0.25	1,476	756	372	1,590	839	432	1,587	836	429
	0.30	1,230	630	310	1,322	697	358	1,320	694	356
	0.35	1,054	540	266	1,130	595	305	1,129	593	304
	0.40	923	473	233	986	519	262	985	517	264
0.70	0.05	6,460	3,300	1,620	7,063	3,739	1,936	7,047	3,723	1,918
	0.10	3,230	1,650	810	3,523	1,863	964	3,515	1,856	956
	0.15	2,153	1,100	540	2,343	1,238	639	2,339	1,233	635
	0.20	1,615	825	405	1,753	926	477	1,749	921	473
	0.25	1,292	660	324	1,399	738	380	1,396	736	378
	0.30	1,077	550	270	1,162	613	315	1,160	610	313
	0.35	923	471	231	994	523	269	992	521	268
	0.40	808	413	203	867	486	234	866	455	232
0.80	0.05	4,920	2,520	1,240	5,450	2,907	1,520	5,483	2,891	1,502
	0.10	2,460	1,260	620	2,718	1,448	756	2,710	1,440	748
	0.15	1,640	840	413	1,807	962	501	1,802	957	496
	0.20	1,230	630	310	1,351	719	374	1,348	715	370
	0.25	984	504	248	1,078	573	298	1,075	569	295
	0.30	820	420	207	896	475	247	894	473	244
	0.35	703	360	177	765	406	210	764	404	208
	0.40	615	315	155	668	353	183	666	353	181
0.90	0.05	2,780	1,420	700	3,185	1,718	918	3,168	1,700	900
	0.10	1,390	710	350	1,587	865	456	1,579	847	447
	0.15	927	473	233	1,054	567	302	1,049	562	296
	0.20	695	355	175	788	423	225	784	419	221
	0.25	556	284	140	628	367	179	625	333	175
	0.30	463	237	117	521	279	148	519	277	146
	0.35	397	203	100	445	238	126	442	236	124
	0.40	348	178	88	308	207	109	386	206	107

Method 1 divides the sample size of diseased subject (n_D), calculated by the approximation normal estimation formula, by the disease prevalence rate (prev). Method 2 includes uncertainty in the prevalence using the approximation normal method to find the smallest number n with probability 0.95 to get equal to or more than n_D disease subjects. Method 3 is similar to Method 2 except exact binomial distribution is used. Since MS Excel 2007 was used to determine the sample size for Methods 2 and 3, there may be small variations compared with other statistical software

investigator can decide how much loss in sensitivity can be borne by applying the new diagnostic procedure. The equivalence does not mean that both tests have equal sensitivities but it is acceptable for a certain range of difference that seems either clinically unimportant or has more gain in term of cost-effectiveness or any other reasons. For example, the sensitivity of MRI in diagnosing breast cancer is

$S_{NO} = 0.90$, whereas another test (CT) to diagnose breast cancer will be accepted if its sensitivity is 0.80 ($\delta_0 = 0.10$) but will be rejected if the sensitivity is only 0.70 (i.e. $\delta_1 = 0.20$). δ_0 is an acceptable difference in the sensitivities. The null and alternative hypotheses are:

$$\begin{aligned} H_0 : &\quad S_{NO} \leq S_N + \delta_0 \\ H_A : &\quad S_{NO} > S_N + \delta_0 => S_N + \delta_1 \end{aligned}$$

δ_1 is the unacceptable difference of two sensitivities $\delta_1 > \delta_0$. Lu and Bean (1995) derived the sample size for the maximum, minimum, and mid-point level of P_{11} in conditional and unconditional formulas. P_{11} is the probability of being diagnosed by both tests. Since P_{11} indicates agreement between two tests with respect to the positivity of the test, of equivalence, there is an inverse relationship between n_D and P_{11} or, in other words, P_{11} is a decreasing function with sample size. A conditional formula is generally recommended for calculation of the sample size. A conditional formula with normal approximation considers disconcordant pairs. The unconditional formula suggested by Conner (1987), which was based on an assumption of conditional independence between the diagnostic tests, in other words, an unconditional formula, does not consider disconcordant pairs. A conditional formula has less sample size than an unconditional formula for similar conditions, which shows that the former method is more powerful than the latter. However, when the probability of discordance is small, the conditional formula is less powerful than the unconditional formula. Lachenbruch (1992) suggests the mid-value approach for choosing P_{11} to determine the sample size instead of the minimum and maximum. The mid-point level of

$$P_{11} = \max\left(S_{NO} - \delta_1 - \frac{(1 - S_{NO})}{2}, \frac{(S_{NO} - \delta_1)}{2}\right)$$

The sample size formula to determine the number of diseased subjects is

$$n_D = \frac{\left\{ \begin{array}{l} Z_{1-\alpha}[2(S_{NO} - P_{11}) - \delta_1]\sqrt{(S_{NO} - P_{11} - \delta_0)} \\ +Z_{1-\beta}[2(S_{NO} - P_{11}) - \delta_0]\sqrt{(S_{NO} - P_{11} - \delta_1)} \end{array} \right\}^2}{(S_{NO} - P_{11})[2(S_{NO} - P_{11}) - \delta_1](\delta_1 - \delta_0)^2}$$

Table 7.5 in the Appendix provides the sample size for one-side equivalence at different sensitivities and various acceptable and unacceptable differences for 80–90 % power and 95 % confidence level.

Example 7 A study is planned to test the one-side equivalence (non-inferiority) of CT in diagnosing breast cancer. The sensitivity of MRI in diagnosing breast cancer is 0.90, whereas a sensitivity of 0.80 is said to be non-inferior. Reject CT if the sensitivity is less than 0.70. Histopathology is considered as the gold standard. Determine the number of subjects required under the conditional formula and mid-point level approach considering 95 % confidence with 80–90 % power.

Table 7.5 Number of subjects required using the mid-point level for one-sided equivalence for various sensitivities with 80 % and 90 % power and 95 % confidence level

		90 % power and 95 % confidence level				80 % power and 95 % confidence level			
δ_0	δ_1	$S_{NO} = 0.95$	$S_{NO} = 0.90$	$S_{NO} = 0.85$	$S_{NO} = 0.80$	$S_{NO} = 0.95$	$S_{NO} = 0.90$	$S_{NO} = 0.85$	$S_{NO} = 0.80$
0.0	0.05	304	488	666	841	226	358	486	612
	0.10	102	152	199	244	78	113	147	179
	0.15	56	80	102	123	44	60	76	91
	0.20	40	51	64	76	29	39	48	57
	0.25	27	37	45	53	22	29	34	40
	0.30	21	28	34	40	17	22	26	30
	0.35	17	23	27	32	14	18	21	24
0.05	0.10	250	412	574	739	190	306	422	541
	0.15	90	135	179	221	70	102	133	164
	0.20	51	73	93	113	41	56	70	84
	0.25	35	48	60	71	28	37	46	54
	0.30	26	35	43	50	21	27	33	38
	0.35	20	26	33	38	17	21	25	29
	0.40	17	22	26	30	14	17	20	23
0.10	0.15	227	372	524	678	175	279	387	498
	0.20	84	125	166	206	66	95	124	153
	0.25	48	67	88	107	39	53	67	80
	0.30	33	45	57	68	27	35	43	51
	0.35	25	33	41	48	20	26	32	37
	0.40	20	25	31	37	16	21	24	28
0.15	0.20	214	351	492	637	166	264	365	469
	0.25	80	118	157	195	63	91	118	146
	0.30	46	65	84	102	37	51	64	77
	0.35	33	43	54	65	26	34	42	49
	0.40	24	32	39	46	20	25	31	36

	0.45	19	24	30	35	16	20	24	27
0.20	0.25	206	336	469	608	161	253	349	449
	0.30	77	114	150	187	62	88	114	140
	0.35	44	63	80	98	37	49	62	74
	0.40	31	42	53	63	26	33	41	48
	0.45	23	31	38	45	20	25	30	35
	0.50	19	24	30	34	16	20	23	27

δ_0 is the acceptable difference between the two sensitivities, δ_1 is the unacceptable difference between two sensitivities

Solution MRI sensitivity $S_{NO} = 0.90$; the acceptable difference between sensitivities $\delta_0 = 0.10$; the unacceptable difference $\delta_1 = 0.20$. The value of P_{11} is max(0.65,0.35). Using Table 7.5, the sample size is 125 subjects with 90 % power and 95 subjects with 80 % power.

Exact Sample Size

The sample size based on the exact binomial distribution is always higher than for the normal approximation. When the expected proportion is closer to 1, the difference in these two methods is substantially significant (Table 7.3). For example, for the sample size for sensitivity of 0.99 with ±0.01 error, and 95 % confidence level, the normal approximation method gives a sample size of 381 subjects; whereas the method based on exact binomial suggests 497 subjects. However, the continuity correction produces a sample size closer to the exact value of 476 subjects.

The Clopper–Pearson method of exact CI estimation has been considered as the standard method and is usually known as the exact method. The sample size in the exact method is obtained by the solving the following equation:

$$\sum_{k=x}^{n_D} \binom{n_D}{k} S_{NL}^k (1 - S_{NL})^{n_D - k} = \frac{\alpha}{2}$$

and

$$\sum_{k=0}^{x} \binom{n_D}{k} S_{NU}^k (1 - S_{NU})^{n_D - k} = \frac{\alpha}{2}$$

where S_{NL} and S_{NU} are the lower and upper limit of the CI of the sensitivities; x and n are the number of true-positives and total number of diseased subjects (Table 7.1). To determine the sample size, fix the lower and upper limit of the CI and alpha, then find n by solving the above equation. Each equation yields a different sample size. The sample size for the study would be the larger of these two sample sizes. For a prospective design, the investigator can apply the prospective design methods as described earlier in the chapter. Fosgate (2005) suggested a modified exact sample size for a binomial proportion and provided sample size tables. Flahault et al. (2005) provide sample size tables for the one-sided exact binomial to test sensitivity with given 5 % alpha and 80 % power.

Software for Determining Sample Size for Sensitivity and Specificity

Almost all the sample size software has provision to calculate the sample size for a case–control design because proportion formulas are applicable. The most frequently used and reliable sample size software packages are NCSS PASS and nQuery. However, prospective design and equivalence for sensitivity and specificity are generally not available in statistical software. PASS-11 uses the naive approach to calculate sample size for sensitivity and specificity in prospective designs for different goals. There are several other programs and calculators available on the Internet. The reliability of this free software is not guaranteed. It is advisable to use NCSS PASS software, which is widely accepted. It is a licensed software but can be downloaded for a free 7-day trial from the following website: http://www.ncss.com. It is difficult to determine the sample size manually using exact methods. These methods need a computer or reliable software

Appendix: How to Use the Goal Seek Command Available in the What if Analysis Tool in the Data Menu for Solving Equations Using MS Excel 2007

To solve the equation:

1. Open an Excel spreadsheet and write the value for case–control sample size obtained (cell B1)
2. Write the value of the sample size obtained using the naive approach in cell B2
3. Write the anticipated prevalence of disease in cell B3
4. Calculate the value of the standard normal distribution using the function NORMSINV for the desired probability; for example, the upper tail for probability 0.95 under the standard normal distribution is -1.645
5. Put the equation $\frac{B1-B3\times B2}{\sqrt{B3\times(1-B3)\times B2}}$ in cell B5

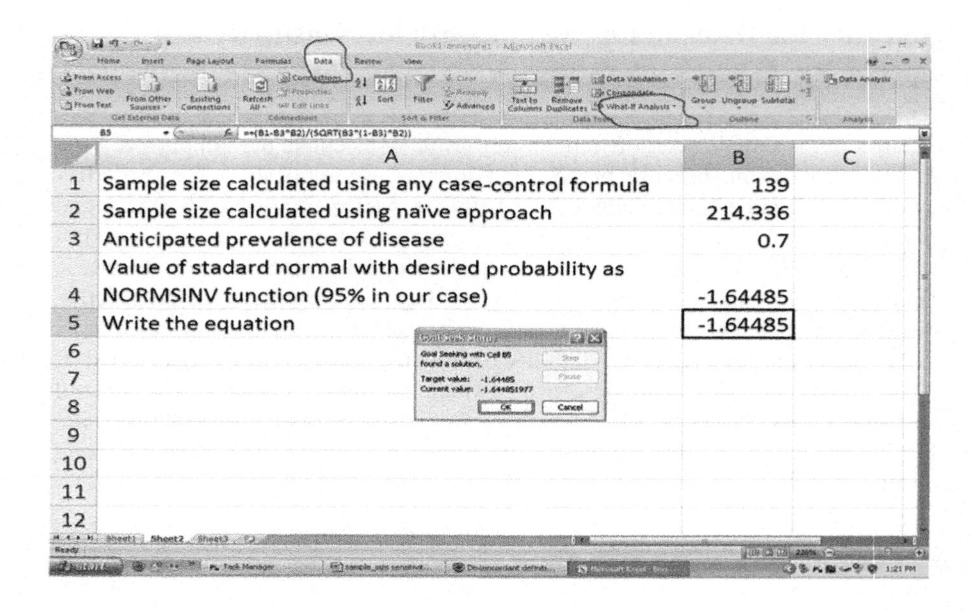

From the Excel menu bar, click Data, locate the Data tool panel and then What if analysis item; from What if analysis select Goal Seek. The Goal Seek dialogue box appears

1. Set cell: insert B5 (the equation as described above)
2. To value: write the value of the standard normal of the preselected probability using the NORMSINV function
3. By changing cell: insert B2

Goal Seek changes the value of cell B2 until the desired solution of the equation is obtained, i.e. −1.64485. The final value of n appears in cell B2.

How to Use the Goal Seek Command to Solve the Exact Binomial Using MS Excel 2007

1. Open an Excel spreadsheet and write the value of the sample size obtained using the naive approach in cell B1 (to avoid convergence problem)
2. Write the value for the case–control sample size obtained (cell B2)
3. Write the anticipated prevalence of disease in cell B3
4. Write the function = 1 − BINOMDIST (x, n, prev, 1) where x is number of successes, n = sample size, Prev is the probability, 1 is the cumulative probability

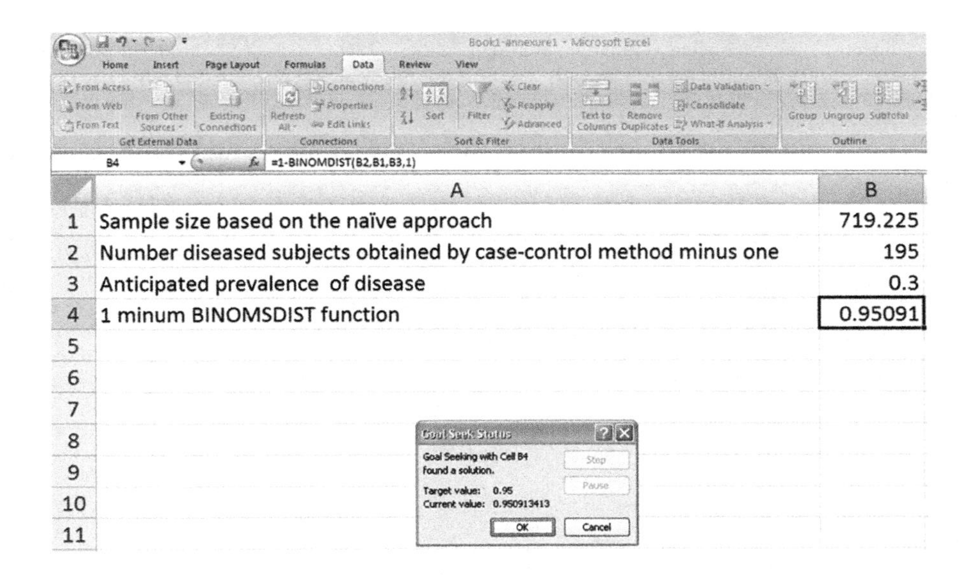

From the Excel menu bar, click Data, locate the Data tool panel and then the What if analysis item; from what if analysis, select Goal Seek. The Goal Seek dialogue box appears

1. Set cell: insert B4, the expression as described above
2. To value: insert the value of the preselected probability = 0.95
3. By changing cell: insert B1

Goal Seek changes the value of cell B1 until 1 – BINOMSDIT is 0.95 and the final value of n appears in cell B1. Sometimes you may not find the exact solution; the user may try other adjacent values of sample size based on the naive approach.

Bibliography

Arkin CF, Wachtel MS (1990) How many patients are necessary to assess test performance? JAMA 263:275–278

Buderer NMF (1996) Statistical methodology: I. Incorporating the prevalence of disease into the sample size calculation for sensitivity and specificity. Acad Emerg Med 3:895–900

Casagrande JT, Pike MC (1978) An improved approximate formula for calculating sample sizes for comparing two binomial distributions. Biometrics 34:483–486

Cohen J (1977) Differences between proportions. In: Statistical power analysis for the behavioral sciences. Academic, Orlando, pp 179–214

Conner RJ (1987) Sample size for testing differences in proportions for the paired sample design. Biometrics 43:207–211

Flahault A, Cadilhac M, Thomas G (2005) Sample size calculation should be performed for design accuracy in diagnostic test studies. J Clin Epidemiol 58:859–862

Fosgate GT (2005) Modified exact sample size for binomial proportion with special emphasis on diagnostic test parameters estimation. Stat Med 24:2857–2866

Hintze J (2011). PASS 11. NCSS, LLC, Kaysville. http://www.ncss.com. Accessed 5 May 2012

Lachenbruch PA (1992) On the sample size for studies based upon McNemar's test. Stat Med 11:1521–1526

Li J, Fine J (2004) On sample size for sensitivity and specificity in prospective diagnostic accuracy studies. Stat Med 23:2537–2550

Lu Y, Bean JA (1995) On the sample size for one-sided equivalence of sensitivities based upon McNemar's test. Stat Med 14:1831–1839

Machin D, Campbell M, Fayers P, Pinol A (1997) Sample size tables for clinical studies, 2nd edn. Blackwell Science, Oxford

Malhotra KR, Indrayan A (2010) A simple nomogram for sample size for estimating sensitivity and specificity of medical test. Indian J Ophthalmol 58:519–522

Schatzkin A, Connor RJ, Taylor PR, Bunnag B (1987) Comparing new and old screening tests when a reference procedure cannot be performed on all screenees. Am J Epidemiol 25:672–678

Chapter 8
An Introduction to Diagnostic Meta-analysis

María Nieves Plana, Víctor Abraira, and Javier Zamora

Abstract Systematic review, and its corresponding statistical analysis, is becoming popular in the literature to assess the diagnostic accuracy of a test. When correctly performed, this research methodology provides fundamental data to inform medical decision making. This chapter reviews key concepts of the meta-analysis of diagnostic test accuracy data, dealing with the particular case in which primary studies report a pair of estimates of sensitivity and specificity. We describe the potential sources of heterogeneity unique to diagnostic test evaluation and we illustrate how to explore this heterogeneity. We distinguish two situations according to the presence or absence of inter-study variability and propose two alternative approaches to the analysis. First, simple methods for statistical pooling are described when accuracy indices of individual studies show a reasonable level of homogeneity. Second, we describe more complex and robust statistical methods that take the paired nature of the accuracy indices and their correlation into account. We end with a description of the analysis of publication bias and enumerate some software tools available to perform the analyses discussed in the chapter.

Introduction

Diagnosis is one of the most prestigious and intellectually appealing clinical tasks among physicians and, usually, the first step in clinical care. Furthermore, because a correct classification of patients according to the presence or absence of a specific clinical condition is essential for both prognosticating and choosing the right treatment, an accurate diagnosis is at the core of high-quality clinical practice. The use of diagnostic tests in clinical practice is generalized. However, introducing

M.N. Plana (✉) • V. Abraira • J. Zamora
Clinical Biostatistics Unit, Hospital Universitario Ramón y Cajal, CIBER en Epidemiología y Salud Pública (CIBERESP) and Instituto Ramón y Cajal de Investigación Sanitaria (IRYCIS), Madrid, Spain
e-mail: nieves.plana@hrc.es

S.A.R. Doi and G.M. Williams (eds.), *Methods of Clinical Epidemiology*, Springer Series on Epidemiology and Public Health,
DOI 10.1007/978-3-642-37131-8_8,

a test into current diagnostic pathways must be preceded by a systematic assessment of its diagnostic performance.

Assessing the value of a diagnostic test is a multi-phase process involving the test's technical characteristics, its feasibility, accuracy, and impact on different dimensions (diagnostic thinking, treatment decisions and, most importantly, impact on patient outcomes). This assessment also includes the social and economic impact of incorporating the test into the diagnostic pathway. Evaluation studies of diagnostic accuracy, in general, and systematic reviews and meta-analyses of studies on test accuracy, in particular, are instrumental in underpinning evidence-based clinical practice. Meta-analysis is a statistical technique that quantitatively combines and summarizes the results of several studies that have previously been included as part of a systematic review of diagnostic tests. A quantitative synthesis of evidence is not always necessary or possible and it is not uncommon to find very high-quality systematic reviews of great informational value for clinical practice that do not include it. Even when a systematic review fails to provide a definite answer regarding the accuracy of a test, it may still contribute valuable information that fills existing scientific gaps and/or informs the design of future primary research studies.

Of the different evaluative dimensions of a diagnostic test, this chapter focuses on test accuracy. Assessing the diagnostic accuracy of a test consists of analysing its ability to differentiate, under the usual circumstances, between individuals presenting with a specific clinical condition (usually a pathology) and those without the condition. For the purpose of this chapter, we assume that diagnostic test results are reported either as positive or negative. This may reflect the actual outcome of the test (e.g. an imaging test result reported as normal or abnormal) or a simplification of a result reported in an ordinal or continuous scale that is then dichotomized into positive/negative using a pre-established cut-off point as with many laboratory results.

In the next section, we revisit the concept of diagnostic accuracy and how it is measured. In the third section, we describe the potential sources of heterogeneity present in systematic reviews of diagnostic test evaluation and how to explore it. The next two sections present two statistical methodologies to choose from according to the presence or absence of inter-study variability. The following section describes publication bias and its analysis. The last section provides a list of software programs available to perform the analyses discussed in the chapter. An appendix with the output of two examples is included.

Evaluation of Diagnostic Accuracy

In contrast with randomized clinical trials where the results regarding the effectiveness of an intervention are reported using a single coefficient (risk ratio, absolute risk reduction, number needed to treat, etc.), individual studies in evaluations of diagnostic test accuracy are summarized using two estimates, which are often inter-related. The statistical methods used to summarize the systematic review results must take into account this dual measurement and report both statistical estimates simultaneously.

As mentioned in Chap. 5 on using and interpreting diagnostic tests, there are several diagnostic accuracy paired measures. These paired estimates are obtained from a 2×2 cross-classification table. The two specific indices conditioned to disease status are sensitivity (the proportion of test positives among people with the disease) and specificity (the proportion of test negatives among people without the disease). Predictive values, positive and negative, are measures conditioned to test results and are calculated as the proportion of diseased individuals among people with a positive test result and the proportion of non-diseased individuals among people with a negative test result, respectively. The well-known impact of the actual disease prevalence on these predictive values discourages their use as summary measures of test accuracy. Likelihood ratios (LRs) are another set of indices obtained directly from sensitivity and specificity. These ratios express how much more likely a specific result is among subjects with disease than among subjects without disease. Another measure of test accuracy is the diagnostic odds ratio (dOR). The dOR expresses how much greater the odds of having the disease are for the people with a positive test result than for the people with a negative test result. It is a single indicator of the diagnostic performance of a test because it combines the other estimates of diagnostic performance in one measure.

Both LRs and dOR index are calculated from the sensitivity and specificity indices and, except under special circumstances, although usually not affected by the disease prevalence, they are affected by the disease spectrum. The dOR index is very useful when comparing the overall diagnostic performance of two tests. Furthermore, because it is easily managed in meta-regression models, it is a valuable tool for analysing the effect of predictor variables on the heterogeneity across studies. However, its use for clinical decision making regarding individual patients is questionable given it is a single summary measure of diagnostic accuracy.

Heterogeneity

Before undertaking a meta-analysis of diagnostic accuracy studies as part of a systematic review, the researcher should ponder the appropriateness and significance of the task. Frequently, the large variability present in sensitivity and specificity indices across the individual studies puts into question the suitability of a statistical pooling of results. A preliminary analysis of the clinical and methodological heterogeneity of the studies should provide the necessary information regarding the appropriateness of synthesizing the results. The selection of potential sources of heterogeneity for further exploration must be done a priori, before starting data analyses, in order to avoid spurious findings. Meta-analysis should only be performed when studies have recruited clinically similar patients and have used comparable experimental and reference tests.

Sources of Heterogeneity

Clinical and Methodological Heterogeneity

In addition to the inherent expected random variability in the results, there can be additional sources of heterogeneity as a result of differences in the study populations (e.g. disease severity, presence of comorbidities), the tests under evaluation (differences in technology or among raters), the reference standards, and the way a study was designed and conducted.

In systematic reviews of treatment interventions, the individual studies usually share a standardized study design (randomized controlled trial or RCT), generally designed with comparable inclusion and exclusion criteria, similar interventions and methods to measure the intervention effect (i.e. similar clinical outcome). In contrast, systematic reviews of diagnostic accuracy studies have to contend with a great deal of variability regarding design, including some studies of questionable methodological quality (retrospective case series, non-consecutive case series, case–control studies, etc.). Empirical evidence shows that the presence of certain methodological shortcomings has a substantial impact on the estimates of diagnostic performance. Pooling results from studies with important methodological shortcomings that have recruited different patient samples may lead to biased or incorrect meta-analysis results.

Threshold Effect

A special source of heterogeneity present in the studies of diagnostic accuracy comes from the existence of a trade-off between sensitivity and specificity. This is a result of the studies using, implicitly or explicitly, different thresholds to determine test positivity.

When studies define different positivity criteria, the sensitivity and specificity change in opposite directions. This effect is known as the threshold effect. As we discuss later, the presence of this effect requires that the meta-analysis consider the correlation between the two indices simultaneously while discouraging analytical strategies based on simple pooling of the sensitivity and specificity measures. Consequently, the meta-analysis of diagnostic accuracy adds a certain level of complexity and requires fitting statistical models, taking into account the covariance between sensitivity and specificity.

Study of Heterogeneity

The first step in a meta-analysis is to obtain diagnostic performance estimates from the individual (or primary) studies included in the review. These data are used to estimate the level of consistency across the different studies (heterogeneity

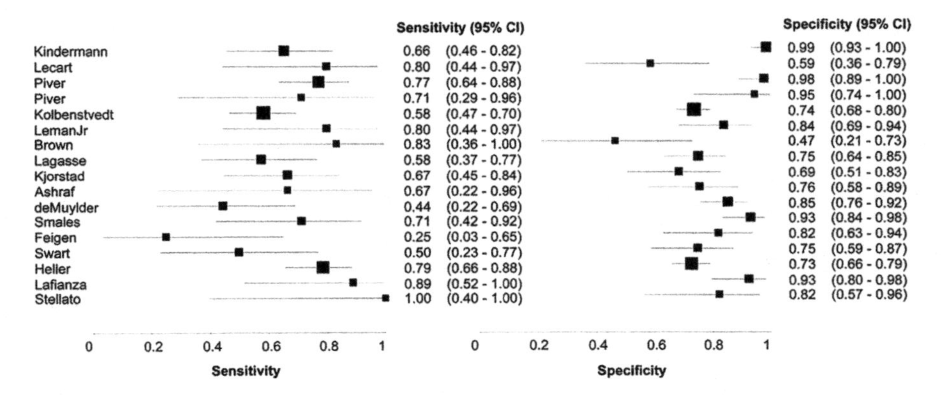

Fig. 8.1 Forest plot of sensitivity and specificity. The box sizes are proportional to the weights assigned and the *horizontal lines* depict the confidence intervals

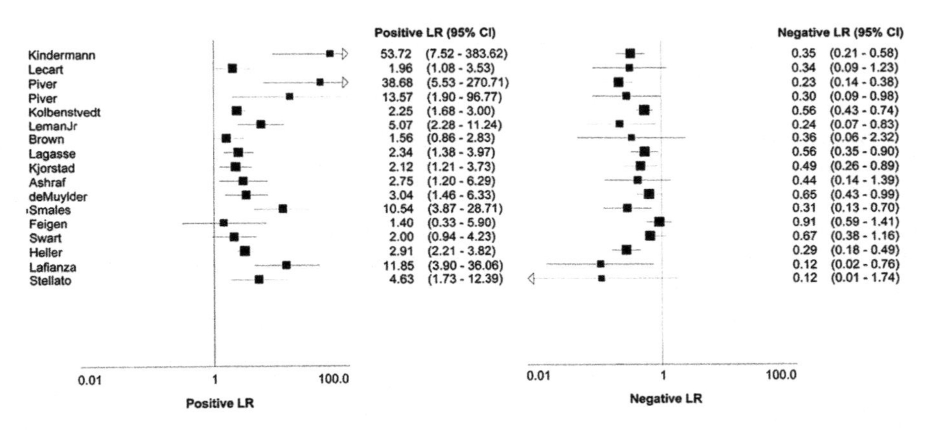

Fig. 8.2 Forest plot of positive and negative LRs. The box sizes are proportional to the weights assigned and the *horizontal lines* depict the confidence intervals

analysis). This description must provide the magnitude and precision of the diagnostic performance indices for every individual study. Given that these accuracy estimates are paired up and are frequently inter-related, it is necessary to report these indices simultaneously (sensitivity and specificity, or positive LR and negative LR). For this description one can use numerical tables of results or paired forest plots (Fig. 8.1) of sensitivity and specificity or of positive and negative LR (Fig. 8.2) for each study together with the corresponding confidence intervals.

A certain level of variability is expected by chance, but the presence of other sources of variation will increase the heterogeneity. These forest plots present the studies ordered from higher to lower sensitivity or specificity (see Fig. 8.4). This format may help analyse consistency among studies and the potential correlation between sensitivity and specificity. However, the best way of illustrating the

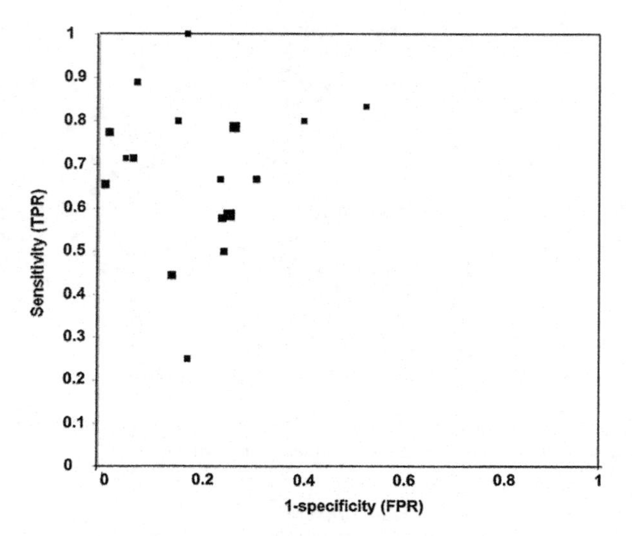

Fig. 8.3 The ROC plane: Plot of 1-specificity against sensitivity

covariance between these indices is to present the pairings of estimates for each study on a receiver operator characteristic (ROC) plot (Fig. 8.3). The *x*-axis of the ROC plot displays the false-positive rate (1-specificity). The *y*-axis shows the corresponding true-positive rate (sensitivity). The higher the diagnostic performance of a study, the closer it is to the upper left quadrant of the ROC plot where both sensitivity and specificity are close to 1. This graphical representation displays a shoulder arm pattern when sensitivity and specificity are correlated, for example, as a result of the presence of a threshold effect or as a result of a different spectrum of the disease among the patients included in the studies. In such situations, sensitivity and specificity are inversely correlated, that is, the true-positive rate (TPR) and the false-positive rate (FPR) are directly correlated.

Specific univariate statistical tests for homogeneity of accuracy estimates have been proposed. However, heterogeneity tests may lack the necessary statistical power to detect heterogeneity when a meta-analysis includes a small number of studies. Conversely, when a meta-analysis includes a large number of studies, heterogeneity tests may detect and interpret slight inter-study variations as strong evidence of heterogeneity by yielding highly significant values, especially when the studies include large sample sizes. In addition, these univariate approaches to heterogeneity analysis do not account for heterogeneity due to the correlation between sensitivity and specificity. The inconsistency index (I^2) may be used to quantify the proportion of total variation across studies beyond what would be expected by chance alone although these estimates must be interpreted with caution.

The results of the heterogeneity analysis must guide the researcher's next step in the completion of the meta-analysis. There are two alternative approaches: (1)

perform separate univariate analyses of the diagnostic accuracy indices; and (2) calculate a pooled estimate of both indices using the appropriate statistical model. Below we describe the two approaches and the circumstances under which one or the other is more appropriate.

Estimate of the Overall Summary Performance of a Diagnostic Test in the Absence of Variability Across Results

The first analytical approach may be used in the special circumstance in which measures of sensitivity, specificity (or both) of the individual studies show a reasonable level of homogeneity. In this scenario, summary estimates of diagnostic accuracy may be obtained through basic meta-analysis techniques with no need for more complex analytical models. Under this approach, two separate poolings of sensitivities and specificities are performed by univariate meta-analysis with fixed or random effects models as deemed appropriate. For added precision, we recommend the use of the logit transformation for sensitivity and specificity to perform the meta-analysis.[1] Once the estimates are averaged, they should be back-transformed to the original scale.

It is important to emphasize that the univariate analysis approach can only be used when there is evidence of homogeneity across estimates. Both sensitivity and specificity indices – and the explicit thresholds defining test positivity, if applicable – must be homogeneous. In this scenario, the correlation between these indices will approach zero and the results of simple pooling will be comparable with those from more advanced models such as bivariate and hierarchical models, discussed later in the chapter. An interesting study concluded that summary indices of diagnostic accuracy calculated with separate simple pooling did not differ significantly from those generated by more statistically robust methods and that the small differences were not clinically relevant.

In the absence of variability across thresholds for test positivity, positive and negative LRs could also be pooled using standard methods such as meta-analysis with fixed or random effects. However, there is some evidence that pooling diagnostic LRs in systematic reviews is not appropriate as the summary LRs generated could correspond to summary sensitivities and specificities outside the valid range from 0 to 1. Instead, it is recommended to calculate the LRs from summary sensitivity and specificity indices estimated using bivariate or hierarchical methods (see below).

We also discourage the practice of averaging predictive values (positive and negative) due to the well-documented effect the prevalence of the disease has on the results. To make matters worse, this prevalence may vary across studies adding an

---

[1] The standard error of a logit transformed proportion p is computed as the square root of $1/(np(1-p))$.

additional source of heterogeneity to the estimates. The summary predictive value is estimated for unknown average disease prevalence. However, in some cases, it is the only index available given the design characteristics of the studies in which reference standards were performed on test positives but not on test negatives (partial verification bias). A typical example of this scenario is when histopathology is used to confirm imaging findings, and no histological sample can be obtained after a negative image result.

Estimate of the Overall Summary Performance of a Diagnostic Test in the Presence of Variability Across Results (sROC Curve)

It is common for researchers performing meta-analyses to run into substantial variability in diagnostic accuracy indices. This second analytical approach addresses the issue of heterogeneity across individual studies. Part of this variability could well originate in differences in the thresholds of positivity used, either explicit or implicit, across studies. Other source of variation could be a differential spectrum of patients across studies. In these cases, separate pooling is not the appropriate method to calculate a summary measure of test accuracy. Instead, the analysis must start by fitting a summary ROC (sROC) curve modelling the relationship between test accuracy measures. Of the different parameters that have been proposed to summarize a sROC curve, the most common is the area under the curve (AUC). This statistic summarizes the diagnostic test performance with only one figure: a perfect test achieves an AUC close to 1, whereas the AUC is near 0.5 for a useless test. This figure may be interpreted as the probability of the test correctly classifying two random individuals, a diseased and a non-diseased subject. Thus, the AUC may be also a useful tool to compare the performance of various diagnostic tests. Another statistic suggested for this task is the Q^* index, defined as the point of the curve in which sensitivity equals specificity. In a symmetric curve, this is the point closest to the upper left corner of the ROC space. The fitted sROC curve may also be used to calculate a sensitivity estimate from a given specificity or vice versa. Two methods for fitting a sROC curve are discussed below.

Moses–Littenberg Model

The Moses–Littenberg method, initially developed to generate sROC curves easily, is the simplest and most popular method for estimating test performance as part of meta-analyses of diagnostic tests. The shape of the ROC curve depends on the underlying distribution of the test results in patients with and without the disease.

There are two methods of fitting the ROC curve. Diagnostic tests where the dOR is constant regardless of the diagnostic threshold have symmetrical curves around the sensitivity = specificity line. When the dOR changes with diagnostic threshold, the ROC curve is asymmetrical. The Moses–Littenberg method is used to study dOR variation according to threshold and thereby generates symmetrical or asymmetrical curves.

The method consists of studying this relationship by fitting the straight line:

$$D = a + bS$$

$$D = \text{logit}(\text{sensitivity}) - \text{logit}(1 - \text{specificity})$$

$$S = \text{logit}(\text{sensitivity}) + \text{logit}(1 - \text{specificity})$$

where D is the natural logarithm (ln) of the diagnostic odds ratio (dOR) and S is a quantity related to the overall proportion of positive test results. S can be considered as a proxy for the test threshold because S will increase as the overall proportion of test positives increases both in the diseased and non-diseased groups. The contrast in test performance variability (measured by dOR) according to threshold is equivalent to the contrast on the model's parameter b. When $b = 0$ there is no variation and the model generates a symmetrical sROC curve; whereas when $b \neq 0$, performance varies according to the threshold and the resulting sROC curve is asymmetrical. The fitting of the previous linear model can be weighted using the inverse variance of ln(dOR) to account for inter-study differences in the sampling error in D.

The model may be expanded to analyse the effect of other factors on diagnostic performance (dOR) as a supplement to the exploration of heterogeneity described here. Such factors, which would be included in the model as covariates, may capture characteristics related to the study design, the patients, or the test.

The Moses–Littenberg model, although very useful for studies of an exploratory nature, is not adequate for drawing statistical inferences. Thus, it should be used keeping in mind some important limitations. First, the model does not take into account either the correlation between sensitivity and specificity or the different precision with which the indices were estimated. In addition, the model's independent variable is random and, thus, its inherent measurement error violates the basic assumption of linear regression models. Finally, the model must be empirically adjusted to avoid empty cells by adding an arbitrary correction factor (0.5).

Bivariate and Hierarchical Models

Two models have been put forward to overcome the limitations ascribed to the Moses–Littenberg model: the bivariate model and the hierarchical sROC model (HSROC). These random effects models are substantially more robust from the statistical point of view than the Moses–Littenberg model. The methodological literature relevant to this area of research proposed these models as the gold standard in meta-analyses of diagnostic accuracy studies. The differences between the two models are small and, in the absence of covariates, both approaches simply amount to different parameterizations of the same model.

The bivariate model is a random effects model based on the assumption that logit (sensitivity) and logit (specificity) follow a normal bivariate distribution. The model allows for the potential correlation between the two indices, manages the different precision of the sensitivity and specificity estimates, and includes an additional source of heterogeneity due to inter-study variance. The second model the methodological literature proposes is known as the hierarchical model or HSROC. It is similar to the previous model except that it explicitly addresses the relationship between sensitivity and specificity using the threshold. Similar to the previous model, this one also accounts for the inter-study heterogeneity.

Both models allow fitting an sROC curve and provide a summary estimate of sensitivity and specificity with the corresponding confidence and prediction intervals. After fitting either of these models, we have to select the most appropriate result to report. It depends on the variability of the results of the individual studies. When sensitivities and specificities of these studies vary substantially, it is advisable to forego average indices and, instead, report the sROC curve. In contrast, when the variability across indices is small, the recommendation is to report the average sensitivity and specificity as calculated based on the bivariate (or the hierarchical) model with its 95 % confidence interval. This is much preferred to the alternative, which would entail risking extrapolating to the ROC space a curve that may misrepresent the test diagnostic accuracy. Summary LRs can be calculated from the pooled estimates of sensitivity and specificity generated by these models. It is worth noting that when an average sensitivity and specificity point is reported over the sROC curve, the position represents the midpoint of the results of the studies calculated based on the average threshold for test positivity, or the average spectrum of the disease, observed in the sample.

Publication Bias

Identifying articles about diagnosis is more cumbersome than finding published clinical trials for a review of intervention performance. Although the MeSH (Medical Subject Heading) term "randomized controlled trial" effectively describes and leads researchers to studies describing clinical interventions, there is no comparable term for the specific literature describing the design of such studies. We should take into account, however, that many studies on diagnosis are based on observations of actual clinical practice in the absence of protocols recorded and/or approved by research ethics committees. For this reason, it is difficult not only to follow up these studies but also to get their results published at the level of detail necessary to be fully useful. If the studies identified in the search were to differ systematically from unpublished manuscripts, the meta-analysis would yield biased estimates that would fail to reflect the real value of diagnostic accuracy.

Similarly, it is also more complex to assess publication bias regarding studies about diagnosis than about treatment. Graphical tools (funnel plots) and the traditional statistical comparisons to evaluate the asymmetry of these graphs were developed to assess publication bias in systematic reviews of clinical trials. Thus, their validity to assess bias in reviews of diagnostic tests is questionable. Deeks and colleagues have adapted the statistical tests of asymmetry of funnel plots to address the issues inherent to meta-analyses of test accuracy. In this version, the funnel plot represents the dOR versus the inverse of the square root of the effective sample size (ESS), which ultimately is a function of the number of diseased and non-diseased individuals. The degree of asymmetry in the plot is statistically evaluated by a regression of the natural logarithm of dOR against $1/ESS^{1/2}$, weighted by ESS.

Software

There is a great variety of statistical packages able to perform the analyses described. Some, like SAS and STATA, are packages for general statistical purposes that facilitate the calculations mentioned by means of a series of macros and user-written commands. The best known user-written commands are the STATA commands MIDAS and METANDI and the SAS macro named METADAS. In addition, the package DiagMeta (http://www.diagmeta.info) was developed for the R environment and it also performs the analyses described.

Additional programs specific to the meta-analysis of diagnostic test accuracy studies are Meta-DiSc and Review Manager (RevMan) by the Cochrane Collaboration. Both perform the basic analyses described in this chapter and RevMan also allows the user to enter parameters obtained from bivariate and hierarchical models and produce corresponding sROC plots.

Appendix

Example 1

For this example we selected the 17 studies included in Scheidler et al.'s meta-analysis (Table 8.1). In their meta-analysis, they evaluated the diagnostic accuracy of lymphangiography (LAG) to detect lymphatic metastasis in patients with cervical cancer.

First, the indices of diagnostic accuracy, sensitivity and specificity (Fig. 8.1) or the positive and negative LRs (Fig. 8.2) of the reviewed studies are described for exploratory purposes using paired forest plots as obtained with Meta-DiSc.

Table 8.1 The studies included in Scheidler et al.'s meta-analysis

id	Study	Year	Test	tp	fp	fn	tn
1	Kindermann	1970	LAG	19	1	10	81
2	Lecart	1971	LAG	8	9	2	13
3	Piver	1971	LAG	41	1	12	49
4	Piver	1973	LAG	5	1	2	18
5	Kolbenstvedt	1975	LAG	45	58	32	165
6	Leman Jr	1975	LAG	8	6	2	32
7	Brown	1979	LAG	5	8	1	7
8	Lagasse	1979	LAG	15	17	11	52
9	Kjorstad	1980	LAG	16	11	8	24
10	Ashraf	1982	LAG	4	8	2	25
11	deMuylder	1984	LAG	8	12	10	70
12	Smales	1986	LAG	10	4	4	55
13	Feigen	1987	LAG	2	5	6	23
14	Swart	1989	LAG	7	10	7	30
15	Heller	1990	LAG	44	50	12	135
16	Lafianza	1990	LAG	8	3	1	37
17	Stellato	1992	LAG	4	3	0	14

From Scheidler et al. (1997)

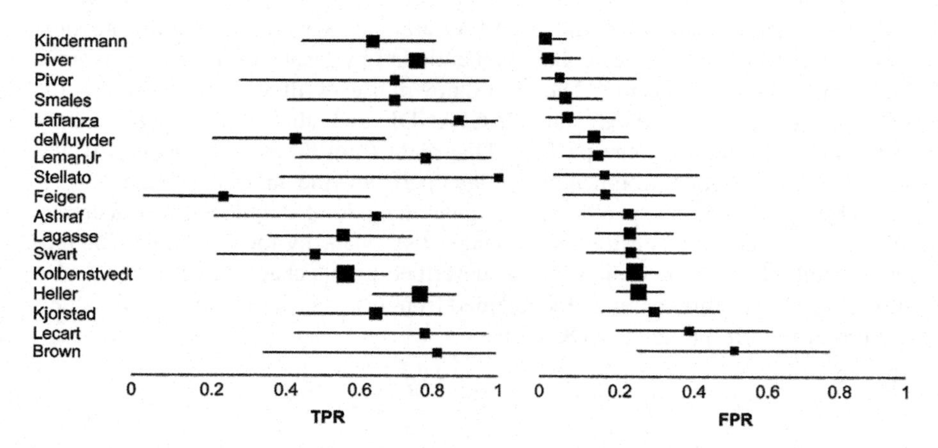

Fig. 8.4 Forest plot with studies sorted by FPR: Heterogeneity is evident

Second, still within the graphical data exploration, we can illustrate the TPR or sensitivity, the FPR (i.e. 1 − specificity), and the LRs (LR + and LR−) organized by one of these indices (Fig. 8.4) or illustrate the pairing indices on a ROC space (Fig. 8.3). At this exploratory phase, all graphical representations should not include pooled estimates of accuracy.

To perform these exploratory analyses, we can use free software (Meta-DiSc, RevMan or the DiagMeta package in the R environment) or any other commercial software.

In this example, and looking at the forest plot generated, we cannot rule out the presence of heterogeneity across the studies included in the review; thus, the analysis should focus on fitting an sROC model.

Given the limitations of the Moses–Littenberg model, we fit a bivariate model using the DiagMeta package. The output is presented below:

> bivarROC(Scheidler)

	ML	MCMC	lower limit	upper limit
average TPR%	**67.38561**	67.59189	60.52091	74.75159
average FPR%	**16.22516**	16.05203	9.25013	25.49491
SD logit TPR	0.34943	0.31889	0.04271	0.87571
SD logit FPR	0.90087	1.06136	0.63934	1.84290
correlation	−0.23882	−0.53898	−0.99999	0.59240

Because the estimated correlation between logit (sensitivity) and logit (specificity) is small and it cannot be ruled out that it is not different from zero, the results estimated by the bivariate model do not significantly differ from those obtained through separate pooling of sensitivity and specificity. Based on the same example, the results using a simple pooling with a fixed or random effects model according to the variability of each of the indices are as follows:

> twouni(subset(Scheidler,GROUP=='LAG'))

TPR	TPR	lower limit	upper limit
Fixed effects	0.6711590	6.218139e-01	0.7169960
Random effects from ML	0.6763973	6.056993e-01	0.7398633
Random effects from MCMC	0.6729242	6.148178e-01	0.7327660
SD of REff	0.0692814	5.935713e-07	0.7516062
FPR	FPR	lower limit	upper limit
Fixed effects	0.1996143	0.1764147	0.2250311
Random effects from ML	0.1619847	0.1059149	0.2397768
Random effects from MCMC	0.1631190	0.1035210	0.2426649
SD of REff	0.9576528	0.5716529	1.5829222

Figure 8.5 shows the sROC curve fitted with a STATA bivariate model, together with the estimated summary point and confidence and prediction intervals.

Example 2

For this illustration we used Fahey et al.'s data (Table 8.2). The goal of their study was to estimate the accuracy of the Papanicolaou (Pap) test for detection of cervical cancer and precancerous lesions.

The sensitivity and specificity forest plots (data not shown) confirm the presence of substantial heterogeneity, in both indices, across the studies included in the review. Figure 8.6 shows the representation of the studies in the ROC space.

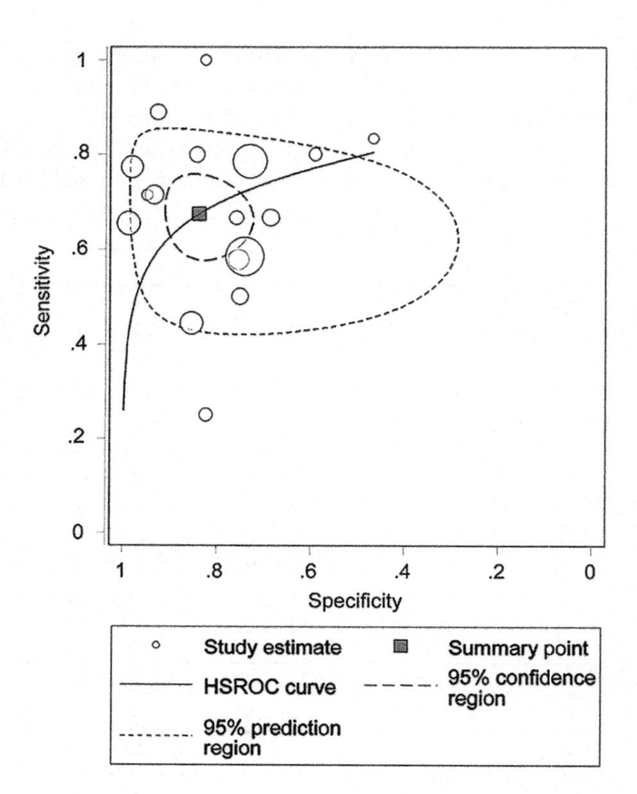

Fig. 8.5 Fitted SROC curve: Study estimates are shown as *circles* sized according to the total number of individuals in each study. Summary sensitivity and specificity are depicted by the *square marker* and the 95 % confidence region for the summary operating point is depicted by the *small oval* in the centre. The larger oval is the 95 % prediction region (confidence region for a forecast of the true sensitivity and specificity in a future study). The summary curve is from the HSROC model

The slight curvilinear pattern of their distribution suggests the presence of a correlation between sensitivity and specificity.

Using Meta-DiSc we calculated the Spearman correlation coefficient between the TPR and FPR logits and obtained a positive and statistically significant correlation of 0.584 ($p < 0.001$) which confirms the results of the bivariate adjustment obtained using the package DiagMeta:

Estimates and 95 % confidence intervals from mcmc samples

	ML	MCMC median	lower limit	upper limit
average TPR%	65.56718	64.93881	57.58497	72.49102
average FPR%	25.38124	25.27866	18.74132	32.57494
SD logit TPR	1.21834	1.27374	1.04000	1.59237
SD logit FPR	1.22834	1.27623	1.02968	1.60834
correlation	0.77408	0.77709	0.61593	0.87730

Posterior probability that rho positive 1
Correlation positive - threshold model appropriate

Table 8.2 Data from Fahy et al.'s study

id	Study	tp	fp	fn	tn	id	Study	tp	Fp	fn	tn
1	Ajons-van K	31	3	43	14	31	Morrison BW	23	50	10	44
2	Alloub	8	3	23	84	32	Morrison EAB	11	1	1	2
3	Anderson 1	70	12	121	25	33	Nyirjesy	65	13	42	13
4	Anderson 2	65	10	6	6	34	Okagaki	1,270	927	263	1,085
5	Anderson 3	20	3	19	4	35	Oyer	223	22	74	83
6	Andrews	35	92	20	156	36	Parker	154	30	20	237
7	August	39	7	111	271	37	Pearlstone	6	2	12	81
8	Bigrigg	567	117	140	157	38	Ramirez	7	4	3	4
9	Bolger	25	37	11	18	39	Reid	12	5	11	60
10	Byrne	38	28	17	37	40	Robertson	348	41	212	103
11	Chomet	45	35	15	48	41	Schauberger	8	4	11	34
12	Engineer	71	87	10	306	42	Shaw	12	2	6	0
13	Fletcher	4	0	36	5	43	Singh	95	9	2	1
14	Frisch	2	2	3	21	44	Skehan	40	18	20	19
15	Giles 1	5	9	3	182	45	Smith	71	13	20	18
16	Giles 2	38	21	7	62	46	Soost	1205	186	454	250
17	Gunderson	4	2	16	31	47	Soutter 1	40	20	17	27
18	Haddad	87	13	12	9	48	Soutter 2	35	9	12	12
19	Hellberg	15	3	65	15	49	Spitzer	10	31	5	32
20	Helmerhorst	41	1	61	29	50	Staff	3	5	3	15
21	Hirschowitz	76	12	11	12	51	Syrjanen	118	40	44	183
22	Jones 1	3	0	5	1	52	Szarewski	13	3	82	17
23	Jones 2	10	4	48	174	53	Tait	38	14	13	62
24	Kashimura 1	28	11	28	77	54	Tawa	16	25	67	291
25	Kashimura 2	79	26	13	182	55	Tay	12	14	6	12
26	Kealy	61	20	27	35	56	Upadhyay	238	52	2	16
27	Kooning 1	62	20	16	49	57	Walker	111	44	20	39
28	Kooning 2	284	31	68	68	58	Wetrich	491	164	250	702
29	Kwikkel	66	25	20	44	59	Wheelock	48	16	38	31
30	Maggi	40	43	12	47						

From Fahey et al. (1995)

With this information in hand, we conclude that the most appropriate method to summarize the results of the meta-analysis is using an sROC curve (Fig. 8.7). This curve was fitted using the bivariate model produced by the macro METANDI in STATA. Figure 8.8 shows the results of a comparable analysis with Meta-DiSc using the Moses–Littenberg model which, in this case, has generated a practically identical sROC curve to that in Fig. 8.7.

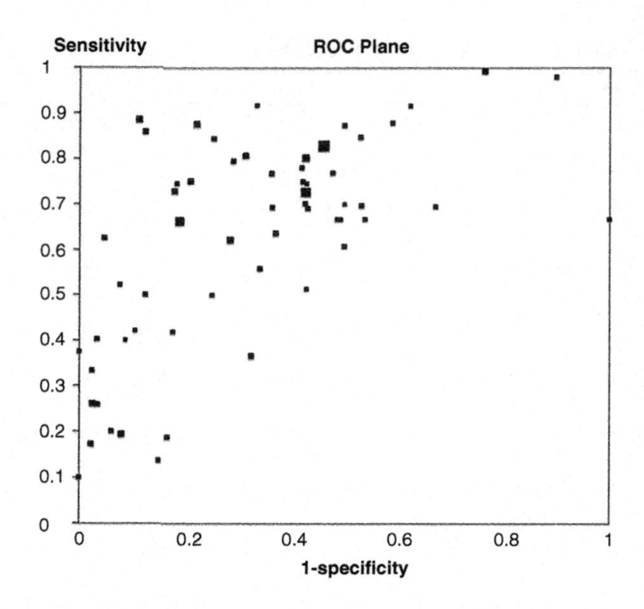

Fig. 8.6 ROC plane: Plot of 1-specificity versus sensitivity

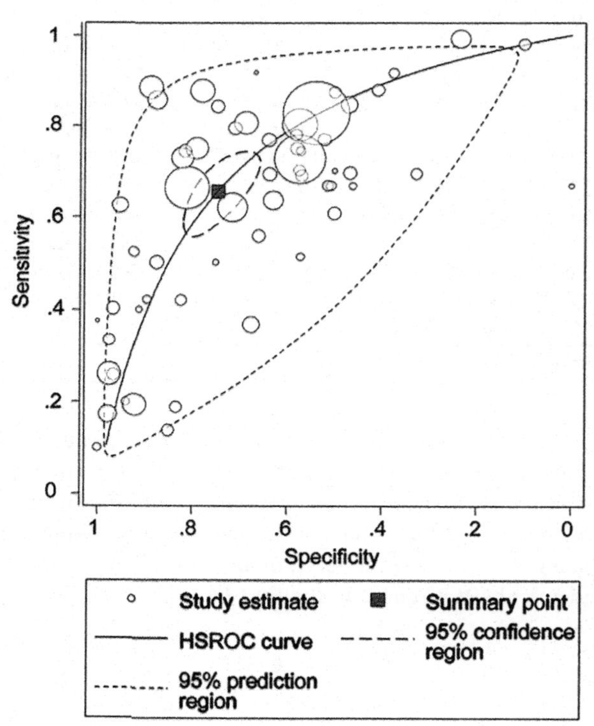

Fig. 8.7 Fitted SROC curve (bivariate model)

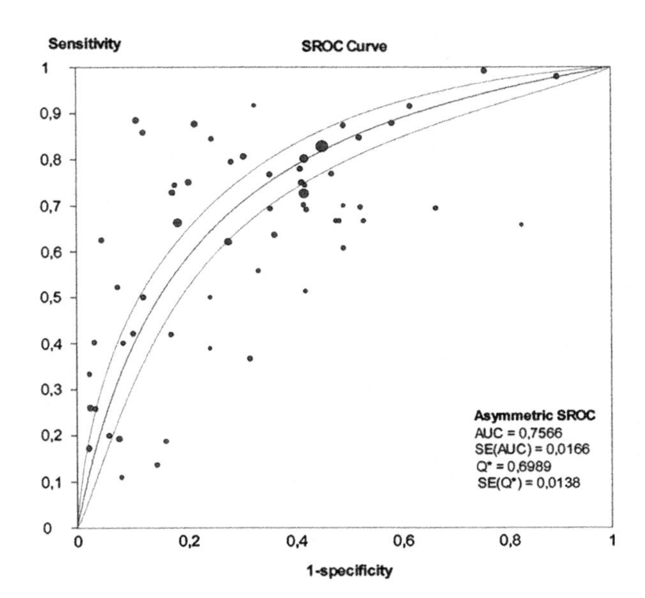

Fig. 8.8 Fitted SROC curve using the Moses–Littenberg model

Bibliography

Begg CB (1994) Publication bias. In: Cooper J, Hedges LV (eds) The handbook of research synthesis. Sage Foundation, New York

Chappell FM, Raab GM, Wardlaw JM (2009) When are summary ROC curves appropriate for diagnostic meta-analyses? Stat Med 28:2653–2668

Deeks JJ, Macaskill P, Irwig L (2005) The performance of tests of publication bias and other sample size effects in systematic reviews of diagnostic test accuracy was assessed. J Clin Epidemiol 58:882–893

Dwamena BA (2007) Midas: a program for meta-analytical Integration of diagnostic accuracy studies in Stata. Division of Nuclear Medicine, Department of Radiology, University of Michigan Medical School, Ann Arbor

Fahey MT, Irwig L, Macaskill P (1995) Meta-analysis of Pap test accuracy. Am J Epidemiol 141:680–689

Gatsonis C, Paliwal P (2006) Meta-analysis of diagnostic and screening test accuracy evaluations: methodologic primer. AJR Am J Roentgenol 187:271–281

Glas AS, Lijmer JG, Prins MH, Bonsel GJ, Bossuyt PM (2003) The diagnostic odds ratio: a single indicator of test performance. J Clin Epidemiol 56:1129–1135

Harbord RM (2008) Metandi: Stata module for meta-analysis of diagnostic accuracy. Statistical software components. Boston College Department of Economics. Chestnut Hill MA, USA

Harbord RM, Deeks JJ, Egger M, Whiting P, Sterne JA (2007) A unification of models for meta-analysis of diagnostic accuracy studies. Biostatistics 8:239–251

Harbord RM, Whiting P, Sterne JA, Egger M, Deeks JJ, Shang A, Bachmann LM (2008) An empirical comparison of methods for meta-analysis of diagnostic accuracy showed hierarchical models are necessary. J Clin Epidemiol 61:1095–1103

Higgins JP, Thompson SG, Deeks JJ, Altman DG (2003) Measuring inconsistency in meta-analyses. BMJ 327:557–560

Honest H, Khan KS (2002) Reporting of measures of accuracy in systematic reviews of diagnostic literature. BMC Health Serv Res 2:4

Lachs MS, Nachamkin I, Edelstein PH, Goldman J, Feinstein AR, Schwartz JS (1992) Spectrum bias in the evaluation of diagnostic tests: lessons from the rapid dipstick test for urinary tract infection. Ann Intern Med 117:135–140

Leeflang MM, Bossuyt PM, Irwig L (2009) Diagnostic test accuracy may vary with prevalence: implications for evidence-based diagnosis. J Clin Epidemiol 62:5–12

Lijmer JG, Mol BW, Heisterkamp S, Bonsel GJ, Prins MH, van der Meulen JH, Bossuyt PM (1999) Empirical evidence of design-related bias in studies of diagnostic tests. JAMA 282:1061–1066

Lijmer JG, Bossuyt PM, Heisterkamp SH (2002) Exploring sources of heterogeneity in systematic reviews of diagnostic tests. Stat Med 21:1525–1537

Lijmer JG, Leeflang M, Bossuyt PMM (2009) Proposals for a phased evaluation of medical tests. Medical tests-white paper series. Agency for Healthcare Research and Quality, Rockville. Bookshelf ID: NBK49467

METADAS (2008) A SAS macro for meta-analysis of diagnostic accuracy studies. User guide version 1.0 beta. http://srdta.cochrane.org/en/clib.html. Accessed 3 July 2009

Moses LE, Shapiro D, Littenberg B (1993) Combining independent studies of a diagnostic test into a summary ROC curve: data-analytic approaches and some additional considerations. Stat Med 12:1293–1316

Reitsma JB, Glas AS, Rutjes AW, Scholten RJ, Bossuyt PM, Zwinderman AH (2005) Bivariate analysis of sensitivity and specificity produces informative summary measures in diagnostic reviews. J Clin Epidemiol 58:982–990

Review Manager (RevMan) (2008) Version 5.0. Copenhagen: The Nordic Cochrane Centre, The Cochrane Collaboration

Rutjes AW, Reitsma JB, Di NM, Smidt N, van Rijn JC, Bossuyt PM (2006) Evidence of bias and variation in diagnostic accuracy studies. CMAJ 174:469–476

Rutter CM, Gatsonis CA (2001) A hierarchical regression approach to meta-analysis of diagnostic test accuracy evaluations. Stat Med 20:2865–2884

Scheidler J, Hricak H, Yu KK, Subak L, Segal MR (1997) Radiological evaluation of lymph node metastases in patients with cervical cancer: a meta-analysis. JAMA 278:1096–1101

Simel DL, Bossuyt PM (2009) Differences between univariate and bivariate models for summarizing diagnostic accuracy may not be large. J Clin Epidemiol 62:1292–1300

Walter SD (2005) The partial area under the summary ROC curve. Stat Med 24:2025–2040

Whiting PF, Sterne JA, Westwood ME, Bachmann LM, Harbord R, Egger M, Deeks JJ (2008) Graphical presentation of diagnostic information. BMC Med Res Methodol 8:20

Zamora J, Abraira V, Muriel A, Khan K, Coomarasamy A (2006) Meta-DiSc: a software for meta-analysis of test accuracy data. BMC Med Res Methodol 6:31

Zwinderman AH, Bossuyt PM (2008) We should not pool diagnostic likelihood ratios in systematic reviews. Stat Med 27:687–697

Chapter 9
Health Technology Assessments of Diagnostic Tests

Bridging the Gap Between Research Findings and Diagnostic Decision Making

Rosmin Esmail

Abstract Evidence-informed decision making with respect to health technology involves considering the best evidence, including the evidence on efficacy, effectiveness and cost-effectiveness of that technology. Health technology assessment (HTA) provides a mechanism to bridge the gap between evidence and decision making through a standard process of formal assessment of this evidence. HTA has been defined by the International Network of Agencies for Health Technology Assessment "as the systematic evaluation of properties, effects, and/or impacts of health care technology, that is, safety, effectiveness, feasibility, cost, cost-effectiveness, and potential social, legal and ethical impact of a technology. It may address the direct, intended consequences of technologies as well as their indirect, unintended consequences". This chapter provides an overview on HTA, its history and methods to conduct a HTA. It also describes how HTAs can be used to assess screening and diagnostic tests to reduce uncertainty and improve appropriateness. A case study on the clinical effectiveness and cost-effectiveness of transcutaneous bilirubin devices for screening and diagnosis of hyperbilirubinemia jaundice illustrates how HTA was applied to assess the evidence on this screening test.

Case Scenario

A newborn baby in Calgary, Alberta, is visually assessed for neonatal jaundice. Based on this assessment, the physician orders the gold standard test for screening neonatal jaundice: a total serum bilirubin (TSB) test. The baby cries and the parents

---

Our stubborn quest for diagnostic certainty ... leads to unnecessary tests supplanting good clinical judgement. Stanley A. Polit, 1989

R. Esmail (✉)
Knowledge Translation, Research Portfolio, Alberta Health Services, Calgary, AB, Canada
e-mail: rosmin.esmail@albertahealthservices.ca

S.A.R. Doi and G.M. Williams (eds.), *Methods of Clinical Epidemiology*, Springer Series on Epidemiology and Public Health,
DOI 10.1007/978-3-642-37131-8_9,

are unhappy that their child has to undergo an invasive test. Reviewing the literature, the physician becomes aware of a new test that is less invasive called transcutaneous bilirubin (TcB) point-of-care testing. For the next newborn baby who may have jaundice, he orders both screening tests TcB and TSB. However, it is soon realized that this test is not available everywhere in Alberta and that there is variation in practice with respect to screening for neonatal jaundice.

There are numerous questions that arise from this scenario. What are some of the considerations that need to be made with respect to the precision and accuracy of the new test (TcB) versus the gold standard (TSB)? Is it ethical to use an invasive test when a less invasive one is available? How efficacious and cost-effective is this new device compared with the gold standard? Do clinicians understand how the new device should be used in screening? What are some of the variations in practice with respect to screening for neonatal jaundice? In trying to ensure an accurate screening and diagnosis, will this lead to ordering unnecessary, costly, and duplicate tests? These are some of the questions that are faced by physicians today when deciding which screening or diagnostic test to use for a particular clinical scenario. Health care organizations and governments may also be interested in similar questions including cost considerations. As stated by Pluddemann et al. (2010), health care purchases and providers have to assess the importance and role that a new diagnostic or screening technology will play in the clinical context.

Evidence-based medicine has been defined as "the conscientious, explicit, and judicious use of the best evidence in making decisions about the care of individual patients" (Sackett et al. 1996). Evidence-informed decision making with respect to health technology involves considering the best evidence including the evidence on efficacy, effectiveness and cost-effectiveness of that technology. When making a decision to fund the technology, therefore, it is vital to have a standard process of formal assessment that can be applied to both screening and diagnostic tests. This chapter focuses on what health technology assessment (HTA) is and how HTAs can be used to assess screening and diagnostic tests to reduce uncertainty and improve appropriateness of use.

History of HTA

The term technology assessment originated in 1965 in the US House of Representatives. In 1973, the Congressional Office of Technology Assessment was founded to review and assess technologies that were used in space. The formal processes for HTA occurred in the mid-1970s with the US Office of Technology Assessment publishing its first report on the subject in 1976. HTA then began to spread around the world with the formation of the Swedish Council on Technology Assessment in Health Care in the late 1980s. Since that time, numerous HTA agencies have been established worldwide including agencies in European countries, Latin America, Asia and Canada. Membership organizations have also been formed including Health Technology Assessment International (HTAi) and the International Network of Agencies for HTA (INAHTA). This is an overarching organization that provides linkage and exchange mechanisms to share results from systematic reviews that agencies undertake. To date, there are 59 members in 29 countries worldwide that are committed to HTA. For further information, Banta and Jonsson (2009) provide an excellent overview of the history of HTA.

What Are Technology, Health Technology, Technology Assessment and HTA?

Definitions of technology, health technology, technology assessment and HTA are described below. The US Congress Office of Technology Assessment defines technology as "the techniques, drugs, equipment, and procedures used by health care professionals in delivering medical care to individuals and the systems within which such care is delivered" (Banta et al. 1981). INAHTA provides a similar definition on health technologies: "prevention and rehabilitation, vaccines, pharmaceuticals and devices, medical and surgical procedures, and the systems within which health is protected and maintained" (International Network of Agencies for Health Technology Assessment 2012).

The US Institute of Medicine's definition of technology assessment is "any process of examining and reporting properties of a medical technology used in health care, such as safety, efficacy, feasibility, and the indications for use, cost, and cost-effectiveness, as well as social, economic, and ethical consequences, whether intended or unintended" (Rettig 1991).

The Canadian Agency for Drugs and Technology in Health (CADTH 2012) defines HTA as "systematically reviewing the research available on technologies with respect to clinical efficacy/effectiveness and/or cost-effectiveness and/or potential service impact. Technologies are defined as drugs, medical devices, medical procedures and health systems used in the maintenance, treatment and promotion of health". The International Network of Agencies for Health Technology Assessment (2012) defines health technology assessment "as the systematic evaluation of properties, effects, and/or impacts of health care technology, that is, safety, effectiveness, feasibility, cost, cost-effectiveness, and potential social, legal and ethical impact of a technology. It may address the direct, intended consequences of technologies as well as their indirect, unintended consequences". Its main purpose is to inform technology-related policy decisions in health care. This is the definition used for the purposes of this chapter. Topics for assessment should be relevant and important to society. The information from a HTA should affect change with respect to making a decision around a policy.

What Are the Steps Involved in Conducting a HTA?

The most recognized source of how to conduct an HTA can be found through Clifford Goodman's text, *HTA 101-Introduction to Health Technology Assessment* published in 2004. Key steps in conducting a HTA are:

1. Identify assessment topics
2. Specify the assessment problem
3. Determine locus of assessment
4. Retrieve evidence
5. Collect new primary data (as appropriate)
6. Appraise/interpret evidence
7. Integrate/synthesize evidence
8. Formulate findings and recommendations
9. Disseminate findings and recommendations
10. Monitor impact

Methodologies that are used to conduct systematic reviews are usually used for steps 1–7 above. For example, the *Cochrane Collaboration Handbook* which has been described in other chapters can be useful in conducting systematic reviews of the literature. In addition, an excellent resource document developed by the Institute for Health Economics entitled, *HTA on the Net* describes various organizations and resources on HTA and is available at http://www.ihe.ca. Step 8 involves making decisions based on the evidence and is usually undertaken by a multidisciplinary team or advisory committee (see next section). Step 9 focuses on the development of an implementation plan and key knowledge management and knowledge transfer strategies to disseminate and share the findings. Step 10 refers to the aspect of whether HTAs are making an impact through monitoring and evaluation.

Another framework that is offered for conducting a HTA is the one by Busse (2002) from the European Collaboration for Health Technology Assessment. The framework is as follows:

- Submission of an assessment request/identification of an assessment need
- Prioritization
- Commissioning
- Conducting the assessment
- Definition of policy question(s)
- Elaboration of HTA protocol
- Collecting background information/determination of the status of the technology
- Definition of the research questions
- Sources of data, appraisal of evidence, and synthesis of evidence for each of:
  - Safety
  - Efficacy/effectiveness
  - Psychological, social, ethical

  - Organizational, professional
  - Economic

- Draft elaboration of discussion, conclusions, and recommendations
- External review
- Publishing of final HTA report and summary report
- Dissemination
- Use of HTA (how effectively were the results adopted and applied?)
- Update of the HTA

Both frameworks have similar steps and either one can be used as a guide to conduct a HTA. Ultimately, what is produced in a HTA is an evaluation of the safety, effectiveness, and cost-effectiveness plus impact of the technology. HTA also considers the social, ethical, environmental, and legal impact of the technology. Therefore, a HTA usually begins with a systematic review (see Chap. 12 for more details) but also considers those aspects that are also part of contextualizing the HTA to the local setting. Unfortunately, many HTAs do not have all of these components because the methodology, which is primarily focused on summarizing qualitative information, has not yet been fully developed.

HTAs also provide the cost-effectiveness of a technology based on economic evaluation. Moreover, economic evaluation is used for determining how we should allocate scarce resources (i.e. each dollar of our budget) to ascertain the best possible outcome (i.e. patient health); HTA within this context helps us to determine the technical and efficient use of resources.

The notion of perspective is also important. A HTA can take on the perspective of the provider, patient, institution or society. Depending on what perspective is taken, the outcomes of the HTA will be different. Therefore, it is important when conducting a HTA that perspective is also considered.

HTA in Canada and Alberta

CADTH "is an independent agency funded by Canadian federal, provincial, and territorial governments to provide health care decision makers with credible, impartial advice and information about the effectiveness and efficiency of drugs and other health technologies" (CADTH 2012). It is a member of INAHTA. HTAs that are produced are context free. HTA reports are prepared by staff, external researchers, or may be collaborative projects between internal and external researchers, both in Canada and internationally. These context-free reports are provided to the organization requesting the report. They are also posted on its website at http://www.cadth.ca. However, these reports need to be contextualized to the setting in which they are going to be used. This involves health technology appraisal or the consideration of context-sensitive issues. Moreover, having systematically reviewed primary and secondary research evidence on whether the technology works and is safe, health technology appraisal addresses the question of will the technology work here in this setting and context and how can it work optimally? It also addresses the issue of whether resources are available to implement the technology and whether additional staff and staff training are required.

Alberta Health Services is an organization of 117,000 staff and 7,400 physicians; it services a population of 3.7 million in the province. In April 2009, the province amalgamated nine health authorities and three provincial entities into one health region that delivers health care services. The Health Technology Assessment and Innovation (HTAI) department was formed in November 2009 (see website: http://www.albertahealthservices.ca/4122.asp). Its mandate is to support teams in managing health technologies through an evidence-informed decision model that helps to:

- Identify, prioritize, and assess health technologies (devices and processes, excluding drugs) expected to significantly affect patient safety, clinical/cost-effectiveness, health outcomes, clinical practice, human resources, and/or policy
- Investigate innovative alternatives for current health technology to improve safety, quality, and/or outcomes
- Promote the effective and appropriate uptake of technologies
- Validate the effectiveness of promising health technologies with access through evidence development initiatives

The HTAI department does not conduct HTAs but can commission HTAs through CADTH and other HTA partners including the Institute for Health Economics and the Universities of Alberta and Calgary. The department also works with the government of Alberta (Alberta Health), and its Advisory Committee on Health Technologies for reviews of health technologies that may be funded at the provincial level. The HTAI department assists in the development of policy by determining what impact this technology will have if introduced into the organization, at the hospital or unit level through operational financial impact assessment (OFIAs). These assessments review the staffing and budget requirements that may be needed to use the technology in practice.

The Role of Regulation of Health Technologies

Regulation also plays a key role in determining whether a particular technology, or diagnostic and screening test in this case, is safe and efficacious. This chapter does not go into detail on how a particular diagnostic or screening test undergoes the testing requirements to ensure safety. Unlike the rigorous testing and clinical trials required for drugs, the same is not true for medical devices. Health Canada does provide licensure for medical devices including screening and diagnostic tests based on efficacy and safety data that are available at the time.

How are HTAs on Diagnostic Tests Prioritized?

HTAs cannot be done on every new diagnostic device that comes on the market or is introduced into the health care system. Moreover, even if the diagnostic or screening test has passed regulatory approval, it may not be known if it is necessarily effective or cost-effective. Pluddemann et al. (2010) has described prioritization criteria to use when assessing new diagnostic technologies based on an expert consensus process. These are described in Fig. 9.1 and can be applied to assess and prioritize diagnostic test as they come to market by health care or other organizations.

Who Uses HTAs?

HTAs can be commissioned and used by:

- Regulatory agencies
- Government and private sector payers
- Managed care organizations
- Health professional organizations
- Hospital and health care networks
- Patient and consumer organizations
- Government, private sector
- Academic health centres
- Insurance companies
- Biomedical research agencies
- Health product companies and venture capital groups

Usually a multidisciplinary working group is formed. This could include the following as members: clinicians, physicians, managers and administrators, laboratory technicians, other technicians, pharmacists, patients/public, epidemiologists, biostatisticians, economists, social scientists, public health and health services researchers, lawyers, ethicists, technology experts, information specialists and members of the public or patients. This working group plays a key role in understanding the issue, determines current practices and formulates the question.

Who Produces HTAs?

The development of a HTA on a particular technology can be completed relatively quickly by commissioning it through several organizations, some of which are listed in Table 9.1. These organizations are usually at arms length, transparent and have individuals skilled in the areas of information sciences, biostatistics,

	Does the Technology Meet the Criterion?		
High Priority	Yes	No	Unsure
1. The potential that the technology will have an impact on **morbidity and/or mortality** of the disease or target condition.			
2. The new technology **reduces the number of people falsely diagnosed** with the disease or target condition.			
3. Improved diagnostic precision using the technology would lead to **improvements in the delivery of treatment** (e.g. shorter time to initiating treatment, reduction in morbidity or mortality).			
4. The new technology improves the ability to **rule out** the disease or target condition.			
5. This disease or target condition to which the diagnostic technology will be applied can be **clearly defined**.			
6. There is evidence of test accuracy in the setting in which the new diagnostic technology will be applied.			
7. The new technology would enhance diagnostic efficiency or be more coot effective than the current diagnostic approach.			
Intermediate Priority			
1. The **prevalence or incidence** of the disease or target condition.			
2. The accuracy of the **current diagnostic approach** tor the disease target condition is problematic.			
3. There is **variation** in treatment or patient outcome resulting from current diagnostic variability.			
4. The current diagnostic pathway for the disease or target condition could be improved by obtaining information in a **less risky fashion** or in a manner more acceptable to patients.			
5. The **safety profile** of the new technology has been established.			
6. The technology improves the ability to **rule in** the disease or target condition.			
7. The new technology has a **clearly defined role** in the diagnostic pathway, e.g. replacing an existing test, as a triage tool, or after the diagnostic pathway as an add-on test.			
8. The relevance of the disease or target condition to current **regional or national health policies and/or priorities.**			
9. It would **be feasible to change** current practice to incorporate this technology (e.g. additional training, infrastructure, or quality control).			

Fig. 9.1 Proposed criteria for the prioritization of diagnostic technologies (Reprinted with permission from Pluddemann et al. (2010))

Table 9.1 International organizations that are involved in HTAs

International HTA societies and agencies	International HTA organizations that conduct HTAs
Health Technology Assessment (HTAi) international: http://www.htai.org	Emergency Care Research Institute: http://www.ecri.org
The International Network of Agencies for Health Technology Assessment: http://www.inahta.org	Canadian Agency for Drugs and Technologies in Health: http://www.cadth.ca
	National Institute for Health and Clinical Excellence: http://www.nice.org.uk/

epidemiology, modelling analysis, economics, etc. They provide the expertise in systematically searching, reviewing and appraising the literature and conducting the economic analysis.

What Is a Diagnostic Test?

Diagnostic and screening technologies are used to provide information that may be used to inform providers whether certain interventions may be used. A diagnostic test can be defined as an information gathering exercise in health care delivery. The purpose of a diagnostic test is to provide information on the presence or absence of a disease. A test can create value in three areas: medical (to inform clinical treatment); planning (to inform patients' choices on reproduction, work, retirement, long-term health, financial plans, etc.); and psychic value (directly changing patients' sense of satisfaction for both positive or negative value). The distribution of value across these dimensions varies from test to test. This in turn affects health outcomes.

Efficacy and Effectiveness of Diagnostic Tests

Health care administrators need to understand and know how health care interventions affect health outcomes. Diagnostic technologies need to demonstrate their efficacy (how well something works in a controlled situation) or effectiveness (how well something works in a population or real setting) to provide information for health care administrators who make decisions on interventions. Goodman (2004) has presented a chain of inquiry that leads from the technical capacity of a technology to changes in patient health outcomes to cost-effectiveness. This is described as follows:

1. Technical aspects: reliable and precise, accurate, operator dependence, feasibility and acceptability, interference and cross-reactivity, inter-and intra-observer reliability
2. Diagnostic accuracy: validity and the "gold standard"
3. Diagnostic thinking
4. Therapeutic effectiveness
5. Patient outcomes
6. Societal outcomes

Fig. 9.2 Important features to determine the usefulness of a diagnostic test (Extracted from Pearl (1999))

1. Technical capacity. Does the technology perform reliably and deliver accurate information?
2. Diagnostic accuracy. Does the technology contribute to making an accurate diagnosis?
3. Diagnosis impact. Do the diagnostic results influence use of other diagnostic technologies? For example, does it replace other diagnostic technologies?
4. Therapeutic impact. Do the diagnostic findings influence the selection and delivery of treatment?
5. Patient outcome. Does use of the diagnostic technology contribute to improved health of the patient?
6. Cost-effectiveness. Does use of the diagnostic technology improve the cost-effectiveness of health care compared with alternative interventions?

Figure 9.2 depicts the important features to determine the usefulness of a diagnostic test. This hierarchy uses six possible end points to determine a test's utility. The criteria are based on Pearl's work (1999). The more criteria in the schema that are fulfilled, the more useful the test; and the less criteria that are fulfilled, the less useful the test.

What Is a Screening Test?

Screening is "the presumptive identification of unrecognized disease or defect by the applications of tests, examinations or other procedures which can be applied rapidly" (US Commission on Chronic Illness 1957). Based on Wilson and Junger's work (1968), a guide to the rational development of a screening program can be based on following these six questions:

- Is there an effective intervention?
- Does intervention earlier than usual improve outcome?
- Is there an effective screening test that recognizes disease earlier than usual?
- Is the test available and acceptable to the target population?
- Is the disease one that commands priority?
- Do the benefits exceed the costs?

Table 9.2 2 × 2 table validating the screening test (*Source*: Bhopal 2002, p. 149)

Screening test	Disease present	Disease absent	Total
Positive	a	b	$a + b$
Negative	c	d	$c + d$
Total	$a + c$	$b + d$	$a + b + c + d$

Sensitivity or true-positive rate = $a/a + c$. Predictive power of a positive test = $a/a + b$
Specificity or true-negative rate = $d/b + d$. Predictive power of a negative test = $d/c + d$

Tests can be used for diagnosis or screening. The basic difference between the two is that diagnosis is conducted on patients who have symptoms, whereas screening is a diagnostic test conducted on patients who do not have symptoms. The goal of all screening tests is to diagnose the disease at a stage that is early enough for it to be cured. This is usually when the patient is asymptomatic and becomes the reason for doing the diagnostic test to validate the screening test. Therefore, the role of screening is to avoid unnecessary diagnostic tests. A screening test aims to have better prognosis (outcome) for individuals; protects society from contagious disease; allows for rational allocation of resources and research.

Other chapters in this book have reviewed the definitions and concepts of using and interpreting diagnostic tests including sensitivity, specificity and likelihood ratios. As with diagnostic tests, sensitivity (or true-positives), specificity (true-negatives) and both positive and negative predictive values also apply to assess the performance of a screening test (Table 9.2).

Reporting and Assessing the Quality of Diagnostic and Screening Tests

Previously, studies on diagnostic and screening tests were not published in a standard format, often leaving readers with many unanswered questions. This led to the development of a set of standards and methods for reporting studies on diagnostic tests, called the Standard for Reporting of Diagnostic Accuracy (STARD). It is a 25-item checklist that can serve as a guide to improve the quality of reporting of a diagnostic study and it has been adopted for preparing journal articles. The Quality Assessment of Diagnostic Accuracy Studies (QUADAS) tool is primarily used to evaluate the quality of a research study that describes a diagnostic test. With respect to screening tests, an article in *JAMA* on "Users guide to the medical literature" by Barratt et al. (1999) provides guidelines on critically appraising studies of screening tests.

Application to Case Scenario

Referring back to the neonatal jaundice example given at the beginning of the chapter, the HTAI department facilitated undertaking a project in spring 2011 that reviewed the clinical and cost-effectiveness of TcB devices for screening and diagnosis of hyperbilirubinemia jaundice. The following case study illustrates how HTA was applied to review this screening test.

Background

Hyperbilirubinemia is the most common cause of neonatal hospital readmissions within Alberta, Canada, and North America. Hyperbilirubinemia jaundice occurs in 60–80 % of normal newborns and nearly all preterm infants. The testing and admission for management of hyperbilirubinemia is a common issue in the care of newborns. It is a complex issue that includes acute care, community providers, primary and tertiary care. It involves a significant burden to the system and requires coordination among these care providers.

Jaundice is due to increased unconjugated and/or conjugated bilirubin levels. Bilirubin is produced by the breakdown of heme-containing proteins including erythrocyte haemoglobin (75 %) and breakdown of other proteins such as myoglobin, cytochromes, catalase, and peroxidases (25 %). Bilirubin must be bound to a form, usually albumin (conjugated), that the body can excrete. Circulating bilirubin that is not bound to albumin is called free bilirubin (unconjugated), which can then enter the brain and cause neonatal injury. Treatment usually involves phototherapy, exchange transfusion for extreme hyperbilirubinemia that is unresponsive to phototherapy, or chemoprevention.

Jaundice is seen in almost all newborns. This is due to a number of factors including increased production of bilirubin (8–10 mg/kg of bilirubin per day); this is twice the amount produced by adults. There is also a decrease in transport and hepatic uptake by the liver, decreased ability to conjugate, and decreased excretion. For these reasons, all newborns experience a rise and then a fall in TSB levels. In normal term infants, bilirubin levels can increase to 5–7 mg/dL found on days 3 and 5 and then decline by day 7–10. These numbers can be different based on gestational age, race and breastfeeding.

Two practice guidelines by the American Academy of Pediatrics have been issued on the management of hyperbilirubinemia and recommend that all infants should be assessed routinely for the development of jaundice (American Academy of Pediatrics Provisional Committee for Quality Improvement and Subcommittee on Hyperbilirubinemia 1994). Furthermore, the Academy of Pediatrics Guidelines (2004) recommends proven prevention strategies for severe neonatal hyperbilirubinemia.

Initially, a visual assessment is conducted in a well-lit room and is performed by blanching the skin of the newborn with slight finger pressure and noting the skin colour. Jaundice is usually visible on the face first, progressing to the trunk and extremities. The next step is to screen for serum bilirubin, which requires measuring the TSB or TcB or a combination of both. Pre-discharge bilirubin screening identifies infants (age 18–48 h) with bilirubin levels >75th percentile for age in hours and tracks those infants with rapid rates of bilirubin, i.e. >0.2 mg/100 ml per hour. These measurements are plotted on an hour-specific nomogram that identifies risk zones and assessment of areas.

TSB remains the gold standard for screening hyperbilirubinemia in newborns. The predictive accuracy of this test has been validated in studies. Non-invasive TcB measurements have been developed for screening newborn infants for hyperbilirubinemia. The BiliCheck™ System and Drager Jaundice Meter JM-103 have been approved by Health Canada and have also been used in the United States. Numerous studies have validated the accuracy of these instruments and values are usually within 2–3 mg/dL of the TSB if the TSB level is less than 15 mg/dL. The accuracy and precision of TcB >15 mg/dL is unproven in comparison with TSB. So, although TcB can be used as a screening tool in the evaluation of hyperbilirubinaemia, it may not replace laboratory measurements of serum bilirubin at higher TSB levels.

Current practices in Alberta for screening and diagnosis of hyperbilirubinemia among newborns are not standardized and vary across zones and urban/rural areas. Most urban areas provide both TcB and TSB, but rural areas are more likely to rely on TSB testing. There is no common nomogram to indicate what TcB threshold should be used for referring a newborn to TSB testing. Practices in communities are even more diversified as there are no guidelines for nurse visits at home or at community clinics to decide under what indications a newborn should be referred to a laboratory for TSB testing, referred to a family physician or hospital. The actual cost for TSB testing is usually less than a dollar for the laboratory assay of a blood sample. There may be additional costs to ensure accuracy. The costs for TcB devices is about ($2,000–4,000) with disposable probes available at a cost of $5 each. There are also additional costs of ensuring and maintaining quality assurance checks of the devices.

Initially an environmental scan was conducted by laboratory services to determine the current practices for testing hyperbilirubinemia in the province. There was no prioritization process conducted by laboratory services to determine if this test indeed should be reviewed. The environmental scan showed that there was no standard practice for TSB/TcB testing in the province and provided the impetus to focus on this topic.

A multidisciplinary team was formed in August 2011 consisting of clinicians, physicians, administrators, laboratory services, contracting and procurement, midwifery, and members of the HTAI team. This team had representation across the province from both urban and rural settings as well as across the continuum of care. It was chaired by laboratory services. Formal terms of reference were developed and meetings were held at least once per month. Assistance on the systematic literature review and cost-effectiveness analysis as part of the HTA was provided by the Institute for Health Economics. The perspective that was undertaken was that from a provider, clinician/provider point of view. Steps in the Busse framework (2004) were used to conduct the HTA.

As stated, HTA is a form of both primary and secondary published research and systematically reviews the existing evidence to inform a policy question. Therefore, it is important to ensure that the policy question is clearly defined. As with a research study or systematic review, usually the PICO methodology is used (i.e. population, intervention, comparator and outcomes). The possible outcomes (effectiveness) of the study were: reducing false-positive and false-negative diagnosis of a newborn with hyperbilirubinemia, reducing readmissions to hospital, and jaundice-related health outcomes.

The following policy question was developed: What is the cost-effectiveness for use of TcB devices for screening and diagnosis of hyperbilirubinemia based on best practices among newborns in Alberta within the first week after birth?

Initially a systematic review of the safety and effectiveness of the screening technology was undertaken. Electronic searches of the literature from 2000 to January 2012 were conducted on the following databases: MEDLINE (including in process), EMBASE, CINAHL, and the Cochrane Database of Systematic Reviews, CRD Databases, and Web of Science. References were also searched. The grey literature was also searched for HTAs, or evidence-based reports, clinical trial registries, clinical practice guidelines, position statutes and regulatory and coverage status. The search was limited to English language articles. Inclusion and exclusion criteria were applied. Title and abstracts were screened and the full text of relevant articles was retrieved. About 40 studies met the inclusion criteria (most were accuracy studies). Theses studies also underwent a quality assessment. A literature review of the economic studies regarding the cost implications or cost-effectiveness of TcB and/or TSB testing was also undertaken. The electronic search covered the period from 2000 to December 2011. The same electronic databases were searched as above including Econlit. References lists within the retrieved articles were also reviewed. There were five studies that met the inclusion/exclusion criteria. Data analysis and synthesis was completed on these. Quality assessment was not conducted. Both literature reviews were completed by June 30, 2012.

The determination of cost-effectiveness for TcB testing is currently underway by the HTAI department. Costs relevant to screening and diagnosis of hyperbilirubinemia are currently being collected including the cost of TcB and TSB tests, quality assurance of TcB and TSB, physician visit, nurse home visit, transportation of blood sample and readmission to hospital. A decision analysis will be developed and cost-effectiveness ratio will be calculated. Together the systematic literature review and cost-effectiveness analysis will inform the practice of using TcB testing in newborns in Alberta. Recommendations from this HTA will be implemented through the development of a standardized provincial clinical pathway for managing neonatal jaundice.

Conclusion

This chapter reviewed what HTA is and how HTAs can be used to assess screening and diagnostic tests to reduce uncertainty and improve appropriateness of use. A case example was shared where an HTA was conducted to inform a policy question.

In our quest for attaining diagnostic certainty, physicians may continue to order more and more tests; some may be useful and others may not. Screening and diagnostic tests are useful if the results of the test changes the course of the treatment. HTA provides a mechanism to evaluate these tests in a rigorous and methodological way so that decisions are informed by evidence and transparent. It considers the safety, efficacy, effectiveness and cost-effectiveness of the technology, so that decisions on the use of a technology can be made with certainty and that the technology is used appropriately. However, it may be that certainty, particularly when it comes to diagnostic and screening tests, is a concept that may not be attainable and as researchers and clinicians we will need to understand and come to terms with this.

Acknowledgements The author would like to acknowledge and sincerely thank Dr. Don Juzwishin, Director, Health Technology Assessment and Innovation, Alberta Health Services; Dr. Mahmood Zarrabi, Senior Health Economist, Health Technology Assessment and Innovation Team, Alberta Health Services; and Ms. Christa Harstall, Director, HTA, Institute for Health Economics, Alberta, Canada, for their editorial feedback and guidance.

Bibliography

American Academy of Pediatrics Provisional Committee for Quality improvement and Subcommittee on Hyperbilirubinemia (1994) Practice parameter: management of hyperbilirubinemia in healthy newborn. Pediatrics 114:297–316

Banta D, Jonsson E (2009) History of HTA: introduction. Int J Health Technol Assess Health Care 25(Suppl 1):1–6

Banta HD, Behney CJ, Willems JS (1981) Toward rational technology in medicine. Springer, New York, p 5

Barratt A, Irwig I, Glasziou P, Cumming RG, Raffle A, Hicks N, Gray JA, Guyatt GH (1999) Users' guides to the medical literature. XVII. How to use guidelines and recommendations about screening. JAMA 281:2029–2034

Bhopal R (2002) Concepts of epidemiology: an integrated introduction to ideas, theories, principles and methods of epidemiology. Oxford University Press, Oxford, pp 145–156

Bhutani VK, Vilms RJ, Hamerman-Johnson L (2010) Universal bilirubin screening for severe neonatal hyperbilirubinemia. J Perinatol 30:S6–S15

Bossuyt PM, Reitsma JB, Bruns DE, Gatsonis CA, Glasziou PP, Irwig LM et al (2003) Towards complete and accurate reporting of studies of diagnostic accuracy: the STARD initiative. Ann Intern Med 138:40–44

Busse (2002) ECHTA Working group 4 Report. http://inahta.episerverhotell.net/upload/HTA_resources/AboutHTA_Best_Practice_in_undetaking_and_reporting_HTAs.pdf

Busse R, Orvain J, Velasco M, Gürtner F, Jørgensen T, Jovell A et al (2002) Best practice in undertaking and reporting health technology assessments. Int J Technol Assess Health Care 18:361–422
Canadian Agency for Drugs and Technology in Health (2003) Guidelines for authors of CADTH health technology assessment reports. http://www.cadth.ca/en/products/methods-and-guidelines. Accessed June 2012
CADTH (2012) Guidelines for authors of CADTH Health Technology Assessment Reports. 16 May 2003, p. 1
Commission on Chronic Illness (1957) Prevention of chronic illness, vol 1, Chronic illness in the U.S. Harvard University Press, Cambridge, p 45
Esmail R (2011) Exploring the relationship between health technology assessment and knowledge management – two sides of the same coin. Canadian College of Health Leaders, Self Directed Learning Paper
Esmail R (2012) Using health technology assessments in health care – are we really making and impact? Canadian College of Health Leaders, Self Directed Learning Paper
Goodman C (2004) HTA 101 introduction to health technology assessment. The Lewin Group, Falls, http://www.nlm.nih.gov/nichsr/hta101/hta101.pdf
Health Technology Assessment and Innovation Department, Alberta Health Services. http://www.albertahealthservices.ca/4122.asp
Institute for Health Economics (2011) Health technology assessment on the net. A guide to internet sources of information. 12th edn. http://www.ihe.ca/documents/HTAontheNet2011.pdf. Accessed June 2012
Institute for Health Economics (2012) Status report: transcutaneous bilirubin testing
International Network of Agencies for Health Technology Assessment (2012) INAHTA Health Technology Assessment (HTA) Glossary. http://www.inahta.org/HTA/Glossary/#_G
Lee DW, Neumann PJ, Rizzo JA (2010) Understanding the medical and nonmedical value of diagnostic testing. Value Health 13:310–314
Lomas J, Culyer T, McCutcheon T, McAuley L, Law S (2005) Conceptualizing and combining evidence for health system guidance. Canadian Health Services Research Foundation, Ottawa
Maisels MJ (2005) Jaundice. In: MacDonald MG, Mullett MD, Seshia MMK (eds) Avery's neonatology: pathophysiology and management of the newborn, 6th edn. Lippincott Williams & Wilkins, Philadelphia, pp 768–846
Maisels MJ, Bhutani VK, Bogen D, Newman TB, Stark AR, Watchko J (2009) Hyperbilirubinemia in the newborn infant ≥35 weeks gestation: an update with clarifications. Pediatrics 214:1193–1198
Mayer D (2004) Essential evidence-based medicine. Cambridge University Press, Cambridge
Norderhaug IN, Sanderberg S, Fossa SD, Forland F, Malde K, Kvinnsland S et al (2003) Health technology assessment and implications for clinical practice: the case of prostate cancer screening. Scan J Clin Lab Invest 63:331–338
Office of Technology Assessment (1976) Development of medical technology, opportunities for assessment. US Government Printing Office, Washington, DC
Pearl WS (1999) Hierarchy of outcomes approach to text assessment. Ann Emerg Med 33:77–84
Pluddemann A, Heneghan C, Thompson M, Roberts N, Summerton N, Linden-Phillips L et al (2010) Prioritisation criteria for the selection of new diagnostic technologies for evaluation. Health Serv Res 10:109
Polit SA (1989) Our stubborn quest for diagnostic certainty. N Engl J Med 321:1272
Rettig RA (1991) Technology assessment – an update. Invest Radiol 26:165–173
Riegelman RK (2005) Studying a study and testing a test. How to read the medical evidence, 5th edn. Lippincott Williams & Wilkins, Philadelphia, p 140
Sackett DL, Rosenberg WM, Gray JAM, Haynes RB, Richardson WS (1996) Evidence-based medicine: what it is and what it isn't. BMJ 312:71–72
Subcommittee on Hyperbilirubinemia (2004) Management of hyperbilirubinemia in the newborn infant 35 or more weeks of gestation. Pediatrics 114:297–316

The Cochrane Collaboration (2011) The cochrane collaboration handbook 5.1. http://www.cochrane.org/training/cochrane-handbook. Accessed June 2012

Trikalinos TA, Chung M, Lau J, Ip S (2004) Systematic review of screening for bilirubin encephalopathy in neonates. Pediatrics 124:1162–1171

Watson RL (2009) Hyperbilirubinemia. Crit Care Nurs Clin North Am 21:97–120

Whiting P, Rutjes AW, Reitsma JB, Bossuyt PM, Kleijnen J (2003) The development of QUADAS: a tool for the quality assessment of studies of diagnostic accuracy included in systematic reviews. BMC Med Res Methodol 3:25

Wilson JMG, Jungner G (1968) Principles and practice of screening for disease. World Health Organization, Geneva

Part III
Modeling Binary and Time-to-Event Outcomes

Chapter 10
Modelling Binary Outcomes

Logistic Regression

Gail M. Williams and Robert Ware

Abstract This chapter introduces regression, a powerful statistical technique applied to the problem of predicting health outcomes from data collected on a set of observed variables. We usually want to identify those variables that contribute to the outcome, either by increasing or decreasing risk, and to quantify these effects. A major task within this framework is to separate out those variables that are independently the most important, after controlling for other associated variables. We do this using a statistical model. We demonstrate the use of logistic regression, a particular form of regression when the health outcome of interest is binary; for example, dead/alive, recovered/not recovered.

The Generalized Linear Model (GLM)

Statistical models are mathematical representations of data, that is, mathematical formulae that relate an outcome to its predictors. An outcome may be a mean (e.g. blood pressure), a risk (e.g. probability of a complication after surgery), or some other measure. The predictors (or explanatory variables) may be quantitative or categorical variables, and may be causes of the outcome (as in smoking causes heart failure) or markers of an outcome (more aggressive treatment may be a marker for more severe disease, which is associated with a poor health outcome).

Generically, a fitted statistical model is represented by linear equations as shown in Fig. 10.1. 'Outcome' is the predicted value of the outcome for an individual who has a particular combination of values for predictors 1–3 etc. The coefficients are estimated from the data and are the quantities we are usually most interested in. The particular value of a predictor for an individual is multiplied by the corresponding coefficient to represent the contribution of that predictor to the outcome. So, in

G.M. Williams (✉) • R. Ware
School of Population Health, University of Queensland, Herston, QLD, Australia
e-mail: g.williams@sph.uq.edu.au

S.A.R. Doi and G.M. Williams (eds.), *Methods of Clinical Epidemiology*,
Springer Series on Epidemiology and Public Health,
DOI 10.1007/978-3-642-37131-8_10,

Fig. 10.1 A fitted GLM depicted mathematically

Predicted outcome = *constant coefficient*
+ *coefficient 1* × value of a predictor 1
+ *coefficient 2* × value of a predictor 2
+ *coefficient 3* × value of a predictor 3
+ ...

particular, if a coefficient for a predictor is estimated to be zero then that predictor makes no contribution to the outcome. The constant coefficient represents the predicted value of the outcome when the values of all of the predictors are zero. This may or may not be of interest or interpretable, because zero may not be in the observable range of the predictor.

So the model predicts values of an outcome from each person's set of values for predictors. This, of course, generally does not match that person's actual observed value. The difference between the observed value and the predicted value is called the residual, or sometimes the error. The term error does not imply a mistake but rather represents the value of a random variable measuring the effects on individual observed outcome values other than those due to the predictor variables included in the model. Adding more predictor variables to the model is expected to reduce the error. Mechanistically, the error or residual for a particular individual is the difference between the individual's observed and predicted values. An example is the difference between an individual's observed blood pressure and that predicted by a model that included age and body mass index.

The theory of model fitting and statistical inference from the model requires that we make an assumption about the distribution of the errors. In many cases, where we have a continuous outcome variable, the assumed distribution is a normal distribution. This is the classic regression model. A log-normal distribution might be used if a continuous variable is positively skewed. However, if we have a binary variable, we might assume a binomial distribution. Thus, the full theoretical specification of a model is represented by Fig. 10.2.

Fitting a Model

Fitting a model means finding the parameter estimates within the model equation that best fits with the observed data. So the parameters referred to in Fig. 10.2 are estimated from the data to give the coefficients referred to in Fig. 10.1. This may be done in different ways. One of the earliest methods proposed to do this was the Method of Least Squares, a general approach to combining observations, developed by the French mathematician Adrien Marie Legendre in 1805. Effectively, this identifies the parameter estimates that minimize the sum of squares of the errors as in Fig. 10.2. In this sense, we estimate the parameters by values that bring the predicted values as close as possible to the observed values. This works well with some probability distributions, but not with others. Currently, the statistically preferred technique is a process called maximum likelihood, or some variant of

Fig. 10.2 The full general linear model depicted mathematically

Outcome = *Constant parameter*
+ *parameter 1* × value of a predictor 1
+ *parameter 2* × value of a predictor 2
+ *parameter 3* × value of a predictor 3
+ ...
+ *Randomly distributed error*

this, which has the advantage of providing a more general framework covering different types of probability distributions. This method was pioneered by the influential English statistician and geneticist, Ronald Fisher, in 1912. The method selects the values of the parameters that would make our observed data more likely (under the chosen probability model) to have occurred than any other sets of values of the parameters. This approach has undergone considerable controversy, application and development, but now underlies modern statistical inference across a range of different situations.

Link Functions

The GLM generalizes linear regression by allowing the linear model to be related to the outcome variable via a link function and incorporating a choice of probability distributions which describes the variance of the outcomes. While this chapter focuses on using the logit link for modelling binary outcomes, it is not the only possible link function. The logit link (hence logistic regression) is linear in the log of the odds of the binary outcome and thus can be transformed to an odds ratio. However, if we want to model probabilities rather than odds, we need to use a log link rather than a logit link and then this can be transformed to a risk ratio. However, unlike the logistic regression model, a log-binomial model can produce predicted values which are negative or exceed one. Another concern is that it is not symmetric since the relative risks for the outcome occurring and the outcome not occurring are not the inverse of each other as with an odds ratio. Also, odds ratios and risk ratios diverge if the outcome is common. If the risk of the outcome occurring is greater than 50 %, it may be better to model the probability that the outcome does not occur to avoid producing predicted values which exceed one.

Models for Prediction Versus Establishing Causality

We can use models to establish causality or for prediction or a combination of the two. For causal models, we are usually interested in ensuring control of confounding, so we can assert that the exposure of interest (say smoking) is a likely cause of the outcome (heart failure); that is, that the association is not due to confounding by social class, diet, etc. In this situation, we usually need to examine closely the relationships between variables in the model. For prediction we try to produce an inclusive model that considers all relevant causes and/or markers of a particular outcome to enable us to predict the outcome in a particular individual. A predictive model thus focuses more on predictor–outcome associations, rather than being concerned with confounding per se.

Now that we have an understanding of a model and its components, we look at a type of model commonly used in clinical epidemiology – logistic regression.

A Preliminary Analysis

Data-set

The Worcester Heart Attack Study examined factors associated with survival after hospital admission for acute myocardial infarction (MI). Data were collected during 13 one-year periods beginning in 1975 and extending until 2001, on all MI patients admitted to hospitals in the Worcester, Massachusetts Standard Metropolitan Statistical Area. The 500 subjects in the data set are based on a 23 % random sample from the cohort in the years 1997, 1999 and 2001 yielding 500 subjects.

Of the 500 patients, 215 (43 %) died within their follow-up period. The median follow-up time was 3.4 years. All patients were followed up for at least 1 year and 138 (27.6 %) died within the first year following the MI. We are interested in examining the factors that predict death within the first year after the MI as the 500 subjects had complete follow-up to this time point.

Preliminary Results

When we examine the risk of death in the first year according to gender and age, we see a somewhat higher percentage of deaths in females than males, and that percentage of deaths increases markedly with age, from 7.2 % (95 % confidence limits (CL) 2.9, 11.6 %) to 49.4 % (41.7, 57.1 %) (Table 10.1). The 95 % confidence intervals are wider for smaller subgroups, but the age variation is substantial. Are these differences statistically significant? Because we are considering two categorical variables, evaluation of statistical significance uses the Pearson chi-square test, provided there are few small expected frequencies. This test examines the null hypothesis that the true risk of death is the same across all subgroups. Implicit in this assertion is an assumption that any observed differences in the estimated risk of death (e.g. 25.0 % vs. 31.5 % for males vs. females) are due to chance. The *P* value associated with the gender comparison is 0.111. Because the *P* value is not small enough (the usual criterion being <0.05), we do not reject our null hypothesis and we conclude the observed differences are not so large that they could not have occurred by chance. For age, however, $P < 0.0001$, and we conclude that observed differences are not consistent with chance variability.

Table 10.1 Percentage of deaths within the first year after an MI, by age group and gender, with 95 % confidence intervals (95 % CI) ($N = 500$)

Risk factor	*N*	Deaths	Deaths (%)	95 % CI	*P* value
Gender					
Male	300	75	25.0	20.1, 29.9	
Female	200	63	31.5	25.1, 37.9	0.111
Age group					
<60 years	138	10	7.2	2.9, 11.6	
60–69 years	86	12	13.9	6.6, 21.3	
70–79 years	114	36	31.6	23.0, 40.1	
>80 years	162	80	49.4	41.7, 57.1	<0.0001

If we are interested in identifying the significance of a trend for risk of death to systematically increase with age, we need to use a statistical test that takes into account how the age categories are ordered. There are various statistical tests and most are available in standard packages. They vary somewhat in their assumptions about the way in which the ordered categories are expressed, but they usually give similar answers, especially in large samples. One of the simplest forms assigns an ordinal score (1,2,3,. . .) to the categories and examines a linear regression of the prevalence on the score (as a predictor variable). For age groups, this test yields a P value < 0.0001.

We can go a step further and examine the relative risks (RRs), that is, the ratio of the percentage of deaths in a subgroup compared with that in a chosen reference group (Table 10.2). In anticipation of later analyses, the odds ratios (ORs) are also given in Table 10.2. Note that ORs are further away from 1 than are relative risks; for example, RR $= 6.81$ for the oldest age group compared with the youngest, with a corresponding OR of 12.49. This will always be the case, and the distance will increase as the risk of death increases. However, this does not change the formal statistical inference regarding this comparison. The P value for the difference between the percentages of deaths is <0.0001, based on a chi-square value of 63.0 (1 df), whether we choose to measure the age effect by an RR, OR, or, indeed a risk difference (49.4 % $-$ 7.2 % $=$ 42.2 %). Table 10.3 shows a similar analysis for selected characteristics of the MI.

We now wish to explore these relationships further to determine which factors, or combinations of them, are the most predictive of death within the first year. We know that the MI characteristics are associated and that they are also likely to be related to age group, itself a strong risk factor. We can explore this in several ways.

One approach is to carry out a stratified analysis: we stratify by a (suspected) confounding variable, and examine the effect of our exposure of interest within each stratum. Thus, to adjust the effect of congestive heart failure for age, we stratify by age groups. Before proceeding further, we collapse age into two categories (<70 years of age and ≥ 70 years) to increase the numbers in each category. Stratified analysis is shown in Table 10.4.

Recall that the RR associated with cardiogenic shock overall was 2.45 (95 % CI 1.73, 3.48) (Table 10.3). We see now that the risk of death is lower in younger

Table 10.2 Percentage of deaths within the first year after an MI, and RRs and ORs by age group and gender, with 95 % CIs ($N = 500$)

Risk factor	N	Deaths (%)	RR	95 % CI for RR	OR	95 % CI for OR
Gender						
Male	300	25.0	1		1	
Female	200	31.5	1.26	0.95, 1.67	1.38	0.93, 2.05
Age group						
<60 years	138	7.2	1		1	
60–69 years	86	13.9	1.92	0.87, 4.26	2.08	0.86, 5.04
70–79 years	114	31.6	4.36	2.26, 8.39	5.91	2.78, 12.57
>80 years	162	49.4	6.81	3.68, 12.63	12.49	6.12, 25.49

Table 10.3 Percentage of deaths within the first year after an MI, and RRs and ORs by MI characteristics, with 95 % CIs ($N = 500$)

Risk factor	N	Deaths (%)	RR	95 % CI for RR	OR	95 % CI for OR
Cardiogenic shock						
Absent	478	25.9	1		1	
Present	22	63.6	2.45	1.73, 3.48	5.00	2.05, 12.20
Congestive heart failure						
Absent	345	17.4	1		1	
Present	155	50.3	2.89	2.19, 3.82	4.81	3.16, 7.32
MI type						
Non-Q wave	347	30.6	1		1	
Q wave	153	20.9	0.68	0.48, 0.97	0.60	0.38, 0.94
History of cardiovascular disease						
Absent	125	24.0	1		1	
Present	375	28.8	1.20	0.85, 1.70	1.28	0.80, 2.04
Atrial fibrillation						
Absent	422	26.1	1		1	
Present	78	35.9	1.38	0.98, 1.93	1.59	0.95, 2.65
Complete heart block						
Absent	489	27.2	1		1	
Present	11	45.4	1.67	0.86, 3.24	2.23	0.67, 7.43
Previous MI						
Absent	329	25.2	1		1	
Present	171	32.2	1.27	0.96, 1.70	1.41	0.94, 2.11

patients (9.8 % vs. 42.0 %). However, within these groups (i.e. controlling for patient age, at least up to a point) we also see that the risk of death increases with the presence of cardiogenic shock, although the RRs have decreased, because of the confounding of the overall effect with age; older patients are more likely to have cardiogenic shock. However, the CIs for these RRs are now wider, reflecting the fact that we are now dealing with subgroups of the data, rather than the entire sample (Table 10.4).

Table 10.4 Percentage of deaths within the first year after an MI, and RRs and ORs by presence of cardiogenic shock and age group, with 95 % CIs

Age	Cardiogenic shock	*N*	Deaths (%)	RR	95 % CI for RR	OR	95 % CI for OR
<70 years	Absent	218	9.2	1		1	
	Present	6	33.3	3.63	1.09, 12.14	4.95	0.85, 28.73
	Total	224	9.8				
≥70 years or more	Absent	260	40.0	1		1	
	Present	16	75.0	1.88	1.36, 2.58	4.50	1.41, 14.33
		276	42.0				

Table 10.5 Percentage of deaths within the first year after an MI, and RRs by presence of cardiogenic shock, with 95 % CIs, unadjusted RR and adjusted by the Mantel–Haenszel method (RR_A) for the effect of age

Cardiogenic shock	*N*	Deaths (%)	RR	95 % CI for RR	RR_A	95 % CI for RR_A
Absent	478	25.9	1	1	1	
Present	22	63.6	2.45	1.73, 3.48	2.02	1.48, 2.76

Using the Mantel–Haenszel technique, we can then pool the stratum-specific RRs, with weightings that reflect stratum size to obtain adjusted RRs. This provides us with the best overall estimate (provided the stratum-specific RRs are consistent) and gives us greater precision, that is, narrow confidence intervals (Table 10.5).

We now see clearly that the effect of adjustment for age appears to have been to decrease the RR associated with cardiogenic shock, since we have adjusted for the fact that patients with cardiogenic shock are also older and age carries its own separate risk.

The Mantel–Haenszel approach to adjustment is an effective method of adjusting for confounders, and is a useful way of identifying confounders one variable at a time. However, it is obvious that this will become tedious when we have multiple confounders to take into account; we would have to construct all the strata related to all combinations of confounder categories, and then perform an analysis on each (some strata would be small, with wide confidence intervals for within-stratum effect estimates) and then pool these estimates. Regression modelling provides us with an effective approach, but, as we will see, involves some additional assumptions.

Logistic Regression

As explained earlier, a regression model consists of two major components: (a) a probability model, which specifies a theoretical distribution (our choice of this is based partly on empirical observations and partly on our theory about the underlying processes that generated the observations) and (b) specification of relevant predictors based on the research questions or hypotheses we want to examine. We now require a probability model for an outcome variable that takes only two values, such as disease/no disease, dead/alive, etc. A further assumption we make is that our observations are independent in the sense that one person dying within the first year after MI does not affect the probability that another person dies in the first year (this may not be true, e.g. if we had included two MI episodes in the study for the same patient). With this assumption, the number of deaths in the first year out of a sample of size N would be expected to follow a binomial distribution. Apart from N, this distribution depends on a parameter p, which is the probability of an event (death in first year). We can estimate this overall by our proportion of deaths, 27.6 % or $p = 0.276$.

However, as we have seen, the risk of death varies according to age and the characteristics of the MI itself. Thus, our p parameter is allowed to take various values, according to various predictors; indeed this is what we want to model. Our outcome variable is the proportion of events of interest (death), out of a given number of possible outcomes, when the probability of a single event is p (which may depend on the predictors of interest).

If we simply model the probability of an event as a function of predictors, it is possible to obtain predicted values that do not lie between 0 and 1. We could, for example, predict a prevalence of −0.05 or −5 % or 1.06 or 106 %. This is a very undesirable feature of a theoretical model.

Several different approaches have been tried to overcome this problem, by transforming the outcome probability to a quantity that must lie between 0 and 1. Currently, the most widely used transformation is the logit transformation, first proposed by Joseph Berkson in 1944. It is effectively a log-odds transformation. If p is the probability of the event of interest (say disease), the logit of p is given by

$$\text{logit}(p) = \log\left(\frac{p}{1-p}\right) = \log(\text{odds of disease})$$

where log is the logarithm function, to base e.

We can see that this transformation accommodates the constraints on modelling a proportion. If we invert the transformation, we can see that the probability of the event, p, is

$$p = \frac{\exp(\text{logit}(p))}{1 + \exp(\text{logit}(p))}$$

where exp is the exponential or antilog function. This is always greater than zero, because the exponential function cannot take negative values. The denominator is larger than the numerator, so it can never be greater than 1. So proportions must lie between 0 and 1.

The main reason for the popularity of this transformation, however, is the consequent interpretation of the regression coefficients when it is used. Putting this transformation together with a model based on age group (reverting to four age groups), we have the logistic regression model as follows:

$$\begin{aligned} \text{logit(Probability of death)} &= \log(\text{odds of death}) \\ &= a + b_1 \times (60 - 69 \text{ years}) \\ &\quad + b_2 \times (70 - 79 \text{ years}) \\ &\quad + b_3 \times (80 \text{ years and over}) \end{aligned}$$

where the notation following the coefficients means: if the statement inside the brackets is true, the value inside the brackets takes the value 1, otherwise it takes the value 0. These are sometimes referred to as indicator variables. This is a compact way of indicating that the coefficients b_1, b_2, b_3 are associated with the categories 60–69 years to 70 years or older, in order, and that the omitted category, <60 years, is the reference category. The above model fits the framework give in Fig. 10.1, where the predicted outcome is the logit(Probability of death), the coefficients are a, b_1, b_2, b_3, and the values of the predictors are given by the indicator variables for each age group.

To further clarify the interpretation of the coefficients, and the role of the reference category, consider a patient who is less than 60 years of age. This patient's predictive model is as follows:

$$\log(\text{odds(death)}), \quad \text{if patient} < 60 \text{ years} = a$$

A patient who is 60–69 years of age has the following predictive model:

$$\log(\text{odds(death)}) \text{ if patient } 60 - 69 \text{ years} = a + b_1$$

Subtracting these last two expressions (the first from the second), we see that

$$\begin{aligned} &\log(\text{odds(death)}) \text{ if patient } 60 - 69 \text{ years} \\ &- \log(\text{odds(death)}) \text{ if patient} < 60 \text{ years} \\ &= b_1 \end{aligned}$$

Using the fact that ($\log A - \log B$) = $\log(A/B)$, we see that

$$\begin{aligned} b_1 &= \log\left(\frac{\text{odds(death) if patient } 60 - 69 \text{ years}}{\text{odds(death) if patient} < 60 \text{ years}}\right) \\ &= \log\left(\begin{matrix}\text{odds ratio of death for patient aged } 60 - 69 \text{ years,} \\ \text{compared with patient aged} < 60 \text{ years}\end{matrix}\right) \end{aligned}$$

So the regression coefficients are directly interpretable as log(ORs) and we can then obtain the actual OR by exponentiation or antilogs of the parameter estimates.

Table 10.6 Parameter estimates from logistic regression of death in the first year, with age group as a predictor

Parameter		Value	Parameter estimate	OR	95 % CL for OR	*P* value
Intercept	a		−2.55			
Age at MI	Reference: Age < 60 years					
	b_1	60–69 years	0.73	2.08	0.86, 5.04	0.106
	b_2	70–79 years	1.78	5.91	2.78, 12.57	<0.0001
	b_3	≥80 years	2.52	12.49	6.12, 25.49	<0.0001

Fitting the logistic regression model is done using maximum likelihood estimation of the model parameters (a, b_1, b_2, b_3 in the age group model), as has been described previously. It is not important to understand the details of this process, but it is important to understand that the process does not always work, in the sense that a solution may not be found, often due to sparseness of data or unusual distributions. Depending on the software you use, you may receive a warning that convergence has not been attained, or you may simply observe results that look meaningless, such as extremely large standard errors of estimates. You should always scrutinize parameter estimates and their standard errors (or CLs) to look for values that differ greatly from your single variable or preliminary analyses.

Again (Table 10.6) we see that the trend for ORs from the logistic regression, as for RRs (Table 10.2), increases as age increases. However, we see that the ORs from the logistic regression are exactly the same as those in Table 10.2. This is because they are mathematically equivalent; this equivalence does not hold as we include more variables in the analysis. The parameters b_1, b_2, b_3 thus represent the outcome (death) log ORs for each group, compared with the reference group, which is the group omitted from the parameter list in the model. The parameter a is usually not of interest; it represents the log(odds) of the event, within the reference category. In this case, the reference category is the youngest age group, and the odds of death for this group is 10/(138 − 10) = 0.078 = $e^{(-2.55)}$.

We can also calculate what our model predicts for the probability of death for each age group by substituting for the parameters a, b_1, b_2, b_3.

$$\begin{aligned}
\text{Probability of death} &= \frac{\exp(\text{logit}(p))}{1+\exp(\text{logit}(p))} \\
&= \frac{\exp(-2.55\ )}{1+\exp(-2.55)} = 0.072 \quad \text{if patient} < 60 \text{ years} \\
&= \frac{\exp(-2.55+0.73)}{1+\exp(-2.55+0.73)} = 0.139 \quad \text{if patient } 60-69 \text{ years} \\
&= \frac{\exp(-2.55+0.1.78\ )}{1+\exp(-2.55+1.78)} = 0.316 \quad \text{if patient } 70-79 \text{ years} \\
&= \frac{\exp(-2.55+2.52\ )}{1+\exp(-2.55+2.52)} = 0.494 \quad \text{if patient} \geq 80 \text{ years}
\end{aligned}$$

We see that the univariable model replicates the observed proportions, which is what we would expect.

Multivariable Logistic Regression

Categorical Predictors

Although univariate logistic regression gives the same results as a simple cross-tabulation, the major advantage of embarking on a logistic regression approach obviously comes from the ability to include additional variables, either as confounders, or as risk factors or predictors in their own right. Later, we also deal with interactions, but for now we examine a logistic regression model that includes age group as a possible confounder to the cardiogenic shock effect. This may be written out exactly as we have done previously, by adding additional terms and regression coefficients to the right-hand side of the model equation:

$$\begin{aligned}&\text{logit(Probability death)}\\&\quad= \log(\text{odds(death)})\\&\quad= a + b_1 \times (60 - 69 \text{ years})\\&\qquad + b_2 \times (70 - 79 \text{ years})\\&\qquad + b_3 \times (\geq 80 \text{ years})\\&\qquad + c \times (\text{cardiogenic shock present})\end{aligned}$$

The maximum likelihood estimates are given in Table 10.7.

The coefficients and ORs for age group have now changed because of the inclusion of an additional variable, cardiogenic shock. They are now the estimated effects, after adjusting (controlling) for the effect of cardiogenic shock. Reciprocally, the effects of cardiogenic shock have been adjusted for age group. To see this, consider a patient who is 60–69 years of age and does not have cardiogenic shock. This patient's predictive model is as follows.

$$\begin{aligned}&\log(\text{odds(death)}), \quad \text{if patient } 60 - 69 \text{ years does not have cardiogenic shock}\\&\quad = a + b_1\end{aligned}$$

A patient who is 60–69 years of age and has cardiogenic shock has the following predictive model:

$$\begin{aligned}&\log(\text{odds(death)}) \text{ if patient } 60 - 69 \text{ years has cardiogenic shock}\\&\quad = a + b_1 + c\end{aligned}$$

Table 10.7 Parameter estimates and 95 % CIs from logistic regression of death, with age and presence of cardiogenic shock

Parameter		Value	Parameter estimate	OR	95 % CL for OR	P value
Intercept	a		−2.59			
Age group	Reference: <60 years					
	b_1	60–69	0.66	1.94	0.79, 4.75	0.147
	b_2	70–79	1.73	5.63	2.63, 12.04	<0.0001
	b_3	≥ 80	2.48	11.98	5.85, 24.55	<0.0001
Cardiogenic shock	Reference: Absent					
	c	Present	1.49	4.46	1.68, 11.82	0.0027

Subtracting these last two expressions (the first from the second), we see that

$$\begin{aligned}&\text{log(odds(death)) if patient } 60-69 \text{ years has cardiogenic shock}\\&-\text{log(odds(death)) if patient } 60-69 \text{ years does not have cardiogenic shock}\\&=c\end{aligned}$$

Using the fact that ($\log A - \log B$) = $\log(A/B)$, we see that

$$\begin{aligned}b_1 &= \log\left(\frac{\text{odds(death) if patient } 60-69 \text{ years has cardiogenic shock}}{\text{odds(death) if patient } 60-69 \text{ years does not have cardiogenic shock}}\right)\\&= \log\left(\begin{array}{l}\text{OR of death associated with having}\\\text{cardiogenic shock in patient aged } 60-69 \text{ years}\end{array}\right)\end{aligned}$$

So we have controlled for age by virtue of holding it constant at 60–69 years. It is easy to see that had we held age constant at some other age group, 70–79 years say, then the same result would have been obtained for the age-adjusted effect of cardiogenic shock. This is an assumption that we make: the effects of variables are constant across values of other variables in the model. This assumption can be relaxed at the cost of making the model more complex; see later section on effect modification.

Returning to the results, we now see similar effects to those we saw with the Mantel–Haenszel analysis for the association between death and cardiogenic shock for age: the effect decreases. We can also see that age is a significant predictor of death. Although these results are consistent with the effects we saw in the Mantel–Haenszel process, they are not the same, largely because ORs are not the same as RRs (except when the outcome rate is very low), but also partly because the method of adjustment by logistic regression is mathematically different from the Mantel–Haenszel approach.

Regression modelling using maximum likelihood fitting also produces likelihood ratio tests, which examine the significance of variables overall. These tests each compare two models: a model that excludes the variable of interest, and one that includes it. The chi-square statistic is a measure of the difference between the

Table 10.8 Likelihood ratio tests for logistic regression of death, with patient age group and presence of cardiogenic shock as predictors

Source	df	Chi-square	*P* value
Patient age group	3	77.70	<0.0001
Cardiogenic shock	1	9.61	0.0019

models and thus can be assessed for statistical significance. These are shown in Table 10.8, and confirm the significance of each of the risk factors independently of the other.

Continuous Predictors

In the above analysis we have grouped age into categories. However, risk increases with increasing age and so it may make sense to treat age as a continuous variable. A simple logistic regression model relating death in the first year to age at MI is then as follows:

$$\begin{aligned}&\text{logit(Probability of death)}\\&\quad=\log(\text{odds of death})\\&\quad=a+b\times\text{Age at MI (years)}\end{aligned}$$

Again we see the meaning of the regression coefficients by considering particular values, say a patient who is 65 years at the MI episode.

$$\begin{aligned}&\text{logit(Probability of death)}_{65}\\&\quad=\log(\text{odds of death})\\&\quad=a+b\times 65\text{ (years)}\end{aligned}$$

Compare this with a patient who is 64 years at the MI episode.

$$\begin{aligned}&\text{logit(Probability of death)}_{64}\\&\quad=\log(\text{odds of death})\\&\quad=a+b\times 64\text{ (years)}\end{aligned}$$

Subtracting these, we have

$$\begin{aligned}&\log(\text{odds of death if patient is 65 years})\\&-\log(\text{odds of death if patient is 64 years})\\&\quad=b\end{aligned}$$

Using the fact that $\log A - \log B = \log(A/B)$, we see that

Table 10.9 Parameter estimates and 95 % CLs from logistic regression of death, with age in years as a continuous predictor

Parameter		Value	Parameter estimate	OR	95 % CL for OR	*P* value
Intercept	*a*		−6.86			
Age (years)	*b*	per year	0.080	1.08	1.06, 1.11	<0.0001

$$b = \log\left(\frac{(\text{odds of death if patient is 65 years})}{(\text{odds of death if patient is 64 years})}\right)$$
$$= \log(\text{odds ratio for death for a 1-year increase in age at MI})$$

Again we see that the regression coefficient is interpretable as a log(OR). Here, however, we do not have a fixed reference group: the OR refers to a fixed increase of 1 unit of the predictor variable. It follows that we cannot interpret the coefficient for a continuous variable unless we know the units in which it is measured. To then get the actual OR we need to exponentiate or antilog the coefficient. Fitting the model for age in years yields Table 10.9.

The OR associated with age is 1.09 or an increase in odds of death by around 9 %. This seems very modest until we remember that this represents the increase associated with only 1 year of age. The predicted increase in risk for an increase of 10 years of age (similar to the age groups we used earlier) can be calculated as follows:

$$\text{Increase in } \log(\text{odds death}) \text{ for 1 year of age} = 0.084$$
$$\text{Increase in } \log(\text{odds death}) \text{ for 10 years of age} = 0.084 \times 10 = 0.84$$
$$\text{Increase in (odds death) for 10 years of age} = e^{0.84} = 2.32$$

Thus, a decade increase in age at MI increases the odds of death in the first year by 2.32-fold.

We need to be extremely careful in interpreting ORs as RRs. It is well known that ORs approximate RRs when the risk of the outcome is small. Small usually means less than about 15 %. The OR is further from 1 than is the RR, as we can see from Tables 10.2 and 10.3. Thus, if the OR is uncritically interpreted as an approximate RR, it will consistently overestimate the strength of the association.

Let us now examine the predictions from our model. Our fitted model (Table 10.9) is

$$\begin{aligned}&\text{logit(Probability of death)}\\&= \log(\text{odds of death})\\&= -6.86 + 0.084 \times \text{Age (years)}\end{aligned}$$

When we do the algebra to express the probability of death in terms of age at MI we get

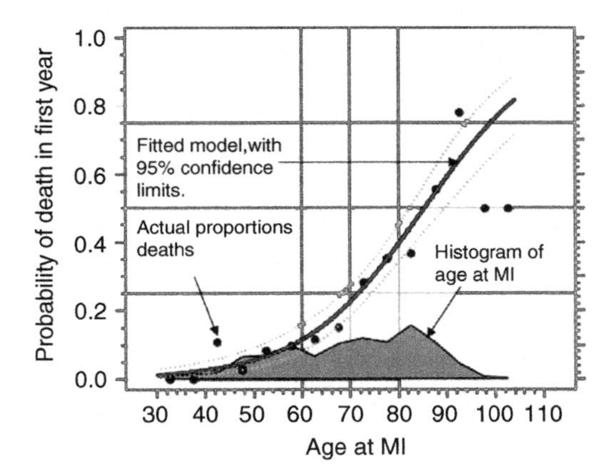

Fig. 10.3 Logistic regression model giving the probability of death in the first year after MI, as predicted by age at MI. The figure shows the classic S-shaped logistic curve; the probability of the outcome increases with the predictor, slowly at first, then increasingly so, and then flattening out. It also shows 95 % CLs for the predicted proportions with the outcome, the risk of the outcome. The dots show the observed risk of death within a centred 5-year age group

$$\text{Probability of death} = \frac{\exp(-6.86 + 0.084 \times \text{Age})}{1 + \exp(-6.86 + 0.084 \times \text{Age})}$$

While this may look a little complex, it is relatively easy to calculate given any particular value of age. Most statistical software programs that fit regression models can calculate these values for all values of predictors that occur in the sample used for fitting the model. We see these in Fig. 10.3 for the current example.

The shaded area at the bottom of the graph in Fig. 10.3 shows the distribution of age, with the three vertical lines showing cut-offs at 60, 70 and 80 years. The RR, comparing two values of the predictor is simply the ratio of the heights of the curve at those outcomes. These values can be read from the graph or calculated from the formula given above. Table 10.10 shows these values, as well as the calculated ORs and RRs, comparing each increase in risk (whether measured by the odds or the proportion of deaths) associated with 1 year increase in age.

Table 10.10 confirms that the OR is constant; this is not surprising because this is a condition of the model. It also confirms that when the predicted probability of death is small (less than 15 %), the RR is very close to the value of the OR. However, as age increases and the predicted risk of death correspondingly increases, the RR diminishes, although it is always >1.

Figure 10.3 is also revealing in terms of the strength of the association between age and death. We see that if a patient is 80 years old or more at the MI, he or she has at least a 50 % chance of dying in the first year after the MI. A patient in the ninth decade of life has an 80 % chance of death in the first year after MI.

Using age as a continuous variable implies that we are fitting a linear effect (on the logit scale) for age; that is, the OR is constant. We may be interested in testing

Table 10.10 Logistic regression model of death with age as a predictor: predicted probabilities, ORs and RRs for each year of age, compared to year below

Age at MI (years)	Predicted probability of death in first year	OR (death) comparing age with age − 1	RR (death) comparing age with age − 1
55	0.08014	1.08	1.08
56	0.08627	1.08	1.08
57	0.09282	1.08	1.08
58	0.09981	1.08	1.08
59	0.10727	1.08	1.07
60	0.11522	1.08	1.07
61	0.12367	1.08	1.07
62	0.13265	1.08	1.07
63	0.14218	1.08	1.07
64	0.15227	1.08	1.07
65	0.16294	1.08	1.07
66	0.17421	1.08	1.07
67	0.18608	1.08	1.07
68	0.19856	1.08	1.07
69	0.21167	1.08	1.07
70	0.22539	1.08	1.06
71	0.23974	1.08	1.06
72	0.25470	1.08	1.06
73	0.27026	1.08	1.06
74	0.28641	1.08	1.06
75	0.30312	1.08	1.06
76	0.32036	1.08	1.06
77	0.33812	1.08	1.06
78	0.35634	1.08	1.05
79	0.37498	1.08	1.05
80	0.39401	1.08	1.05
81	0.41336	1.08	1.05
82	0.43298	1.08	1.05
83	0.45282	1.08	1.05
84	0.47280	1.08	1.04

whether this is a reasonable fit to the data. We can do this by including a square or quadratic term in the model. It is usually helpful to centre continuous variables before including them in polynomial or interaction terms. Centring means subtracting a central value (mean or median) from each value. When we do this we obtain Table 10.11.

We see that the quadratic term is clearly non-significant, indicating the linearity assumption is supported.

Table 10.11 Logistic regression model of death with age as a predictor and a quadratic term

Parameter		Value	Parameter estimate	OR	95 % CL for OR	*P* value
Intercept	*a*		−1.26			
Age at MI						
Age − 70	*b*	per year	0.08	1.08	1.06, 1.10	<0.0001
(Age − 70)^2	*c*	per (year)^2	0.0002	1.00	1.00, 1.00	0.772

Combining Categorical and Continuous Predictors

We can combine categorical and continuous predictors in a model provided we keep in mind the appropriate interpretation of the regression coefficients. We now add the effect age as a continuous variable to a model incorporating cardiogenic shock and gender (both categorical variables), as follows:

$$\begin{aligned}&\text{logit(Probability of death)}\\&\quad= \log(\text{odds of death})\\&\quad= a + b \times \text{Age at MI (years)}\\&\qquad + c \times (\text{Patient is male})\\&\qquad + d \times (\text{Cardiogenic shock is present})\end{aligned}$$

The maximum likelihood estimates of the model parameters are now given in Table 10.12.

The inclusion of age as a continuous variable and gender has reduced the effect of cardiogenic shock as a predictor of death, but only slightly. Although females have a higher odds of death than males, this was not significant, and it is likely that the adjustment to the effect of cardiogenic shock was largely due to the strong effect of age, which appears unaffected by adjusting for gender and cardiogenic shock.

Likelihood ratios tests also show the overall significance of effects (Table 10.13), and confirm the predominance of the age and cardiogenic shock effects.

Effect Modification

The models considered so far assume that the effects of predictors are additive on a logit scale; there is only one parameter for the effect of cardiogenic shock, for example, and its effects are assumed to be the same over all age groups. If we wish to allow for effects to vary across values of another variable we need to incorporate an interaction term, which allows for effect modification.

To see how this works, consider the effect of congestive heart failure stratified by age group. Again, for simplicity we divide age into two groups: <70 years and ≥70 years. The stratified analysis is given in Table 10.14.

Table 10.12 Logistic regression model of death, with patient age (as a continuous variable) and gender, and presence of cardiogenic shock

Parameter	Value	Parameter estimate	OR	95 % CL for OR	*P* value
Intercept	a	−7.15			
Age					
	b Years	0.08	1.09	1.06, 1.11	<0.0001
Gender	Reference: Males				
	c Females	0.19	1.21	0.77, 1.90	0.404
Cardiogenic shock	Reference: Absent				
	d Present	1.46	4.29	1.62, 11.33	0.003

Table 10.13 Likelihood ratio tests for logistic regression model of death with patient age (as a continuous variable) and gender, and presence of cardiogenic shock

Source	df	Chi-square	*P* value
Age (years)	1	85.2	<0.0001
Gender	1	0.70	0.403
Cardiogenic shocks	1	9.13	0.0025

Table 10.14 Percentage of deaths within first year after an MI, and RRs and ORs by presence of congestive heart failure and age group, with 95 % confidence intervals

Age	Congestive heart failure	*N*	Deaths (%)	RR	95 % CI for RR	OR	95 % CI for OR
<70 years	Absent	181	5.0	1		1	
	Present	43	30.2	6.08	2.78, 13.30	8.28	2.94, 23.75
	Total	224	9.8				
≥70 years	Absent	164	31.1	1		1	
	Present	112	58.0	1.87	1.41, 2.46	3.06	1.80, 5.21
		276	42.0				

We see that the effect of congestive heart failure is much greater in those who are <70 years of age. Note again that the ORs are further away from 1 than are the RRs. The logistic regression model incorporating age group and congestive heart failure is:

$$\begin{aligned}&\text{logit(Probability of death)} = \text{log(odds of death)}\\&\quad = a + b \times (\text{Age} \geq 70 \text{ years}) + c \times (\text{Congestive heart failure is present})\end{aligned}$$

If we fit this logistic regression (first without allowing for an interaction), we get the results in Table 10.15. The OR for the association of age and death is 5.52 and 95 % CI (3.29, 9.24). The OR for the association of congestive heart failure and death is 3.85 (2.47, 6.01). We see that the logistic regression estimate for congestive heart failure falls between the two age stratum-specific estimates in Table 10.15. Thus, the model averages in some way over the stratum-specific estimates, as it has only one parameter.

Table 10.15 Parameter estimates, ORs and 95 % CIs from logistic regression of death, with age and presence of congestive heart failure

Parameter	Value	Parameter estimate	OR	95 % CL for OR	*P* value
Intercept	*a*	−2.60			
Age group	Reference: <70 years				
	b ≥70 years	1.71	5.52	3.29, 9.24	<0.0001
Congestive heart failure	Reference: Absent				
	c Present	1.35	3.85	2.47, 6.01	<0.0001

The next step is to estimate effects of congestive heart failure within age groups. This is achieved in logistic regression by including additional terms in the predictor part of the model. These additional parameters allow an increment to the congestive heart failure effect for the older age group compared with the younger age group, These parameters are denoted by the *d* parameter in the following formula:

$$
\begin{aligned}
&\text{logit(Probability of death)} \\
&\quad = \log(\text{odds of death}) \\
&\quad = a + b \times (\text{Age } \geq 70 \text{ years}) \\
&\qquad + c \times (\text{Congestive heart failure is present}) \\
&\qquad + d \times (\text{Age } \geq 70 \text{ years}) \times (\text{Congestive heart failure is present})
\end{aligned}
$$

After maximum likelihood fitting of the interaction model we have the results in Table 10.16.

Table 10.16 shows that the interaction parameter *d* falls just short of significance, although as it is very close, we may still be interested in reporting the result. We need to take care in interpreting the above parameter estimates. The antilog of the *c* parameter for age group (e^c) is the OR for those with congestive heart failure compared with those without, within the reference category for age (patients <70 years). It does not represent the overall effect of congestive heart failure (indeed we have assumed there is no overall effect, because it is modified by age). To get the estimated OR for congestive heart failure for those 70 years of age, we add the parameters *c* and *d* together and then antilog to obtain 3.06. Equivalently we can multiply the OR associated with the reference category for age (8.28) by the OR calculated for the interaction parameter (0.37). We usually present model output involving an interaction as in Table 10.17. This table shows the separate ORs for each age group explicitly (which Table 10.16 does not), and the results of the test for interaction. Notice that no overall effects are given for variables involved in the interaction.

As a final example, Table 10.18 displays a model combining cardiogenic shock, age group and congestive heart failure, incorporating the effect modification of congestive heart failure by age group. To demonstrate the parameterization of the model, the model equation is given below.

Table 10.16 Parameter estimates, ORs and 95 % CIs from logistic regression of death, with age and presence of congestive heart failure (CHF) and interaction effects

Parameter		Value	Parameter estimate	OR	95 % CL for OR	*P* value
Intercept	*a*		−2.95			
Age group		Reference: <70 years				
	b	≥70 years	2.15	8.63	4.09, 18.21	<0.0001
CHF		Reference: Absent				
	c	Present	2.11	8.28	3.25, 21.08	<0.0001
Age × CHF	*d*	Reference: Age < 70 years or CHF absent				
		Age < 70 years and CHF present	−0.99	0.37	0.13, 1.07	0.066

Table 10.17 Logistic regression model of death within first year after an MI with age group and presence of congestive heart failure as predictors, allowing for effect modification

Parameter		Value	Parameter estimate	OR	95 % CL for OR	*P* value
Age group						
<70 years						
Congestive heart failure		Reference category: absent				
	c	Present	2.11	8.28	3.25, 21.08	<0.0001
≥ 70 years						
Congestive heart failure		Reference category: absent				
	c + *d*	Present	1.12	3.06	1.86, 5.05	<0.0001
Age × Congestive heart failure		Reference category: <70 years. No congestive heart failure				
≥70 years, No congestive heart failure	*d*		2.15	8.63	4.09, 18.21	<0.0001

$$
\begin{aligned}
&\text{logit(Probability of death)} \\
&\quad = \log(\text{odds of death}) \\
&\quad = a + b \times (\text{Cardiogenic shock is present}) \\
&\qquad + c \times (\text{Age} \geq 70 \text{ years}) \\
&\qquad + d \times (\text{Congestive heart failure is present}) \\
&\qquad + e \times (\text{Age} \geq 70 \text{ years}) \times (\text{Congestive heart failure is present})
\end{aligned}
$$

The way in which the effect of cardiogenic shock is presented has not changed, because it is not involved in an interaction. However, its value has reduced somewhat from its previous value (Table 10.7). This is because of the additional adjustment for congestive heart failure. In the presence of an interaction in the model, other coefficients will be adjusted for all combinations of the interacting variable (equivalent in this case to stratifying by age group and congestive heart failure simultaneously (four groups) and examining the cardiogenic shock effect within each).

Table 10.18 Logistic regression model of death within first year after an MI with cardiogenic shock, age group and presence of congestive heart failure as predictors, allowing for effect modification

Parameter		Value	Parameter estimate	OR	95 % CL for OR	*P* value
Cardiogenic shock		Reference category: absent				
	b	Present	1.27	3.57	1.26, 10.10	0.016
Age group						
<70 years						
Congestive heart failure		Reference category: absent				
	d	Present	2.10	8.13	3.18, 20.81	<0.0001
$\geq$ 70 years						
Congestive heart failure		Reference category: absent				
	$d + e$	Present	1.04	2.83	1.70, 4.70	<0.0001
Age $\times$ Congestive heart failure		Reference category: <70 years, No congestive heart failure				
$\geq$ 70 years, No congestive heart failure	e		2.17	8.75	4.13, 18.53	<0.0001

Likelihood ratio tests are available for each of the terms in our model. For the model in Table 10.18 these are given in Table 10.19.

P values for likelihood ratio tests in Table 10.19 are slightly different from those for parameter estimates given in Table 10.18; for example, the *P* value for the interaction term is $P = 0.066$ in Table 10.18 and 0.051 in Table 10.19. This is because these are estimated in different ways. The likelihood ratio tests are based on the likelihood function for the interaction model compared with the non-interaction model, whereas the *P* values for individual parameters are based on Wald statistics, which relate to the parameter estimates themselves and their standard errors. The likelihood ratio test is generally preferred for various statistical reasons, but both usually give similar answers. It is important to remember that calculation of each of these and indeed many *P* values is an approximate process that relies on large enough sample sizes and is based on assumptions that are sometimes slightly different.

Extensions and Variations of Logistic Regression

Case–Control Studies

Case–control studies address questions of associations between risk factors, commonly called exposures, and health outcomes. Typically a series of cases is first defined. These are persons experiencing the event of interest, for example, successful recovery from an illness. A series of controls is then chosen, according to criteria such that a selected control would have become a case, had he or she had the

Table 10.19 Likelihood ratio tests for logistic regression model of obesity, at the 21-year follow-up, with maternal smoking and child's exercise at age 14 years as predictors, with interaction effects

Source	df	Chi-square	*P* value
Cardiogenic shock	1	6.21	0.0127
Age group	1	44.12	<0.0001
Congestive heart failure	1	19.04	<0.0001
Age × Congestive heart failure	1	3.81	0.051

particular health outcome of interest. An example might be a series of patients experiencing a nosocomial infection during a hospital stay, with controls being chosen from other in-patients who did not experience an infection. In such a case, the variable indicating caseness (case/control) is used as the outcome variable and potential risk factors are included in the logistic regression model in the usual way. If the controls are matched in some way to the cases (e.g. by age, type of ward, admission diagnosis) then a technique called conditional logistic regression is needed to take the matching into account.

Multinomial and Ordinal Regression

The logistic regression model can be extended to the situation when the outcome variable has more than two categories (multinomial regression) and when these categories fall into a natural order (ordinal regression). These models are very similar to the logistic regression but allow the incorporation of additional hypotheses concerning these additional categories of outcome. In many instances it is possible to address the questions dealt with by these more complex models, by using a series of simpler logistic regressions.

Conclusion

Logistic regression is a very general model that can be used to analyse the determinants or predictors of a binary outcome arising in a process in which events are independent. Because of the nature of the logit transformation, the model gives rise to regression coefficients that are interpretable as log(ORs), which allows a useful interpretation, after exponentiation.

As with other regression models, multivariate models can be built up by including additional predictor variables, such that effects are mutually adjusted.

Logistic regression may be applied to continuous variables, or a mix of continuous and categorical variables. Detailed examination of relationships with continuous variables may be valuable in detecting curvilinear effects.

Caution must be exercised in interpreting ORs as RRs. When the outcome becomes more common (at least 15 %), this interpretation may be misleading.

Bibliography

Hosmer DW, Lemeshow S (2000) Applied logistic regression. Wiley, New York

Kirkwood BR, Sterne JAC (2003) Essential medical statistics, 2nd edn. Blackwell Science, Malden, Part C

Kleinbaum DG, Kupper LK, Muller KE (1988) Applied regression analysis and other multivariable methods, 2nd edn. PWS-Kent, Boston, Chapters 21–25

Tonne C, Schwartz J, Mittleman M, Melly S, Suh H, Goldberg R (2005) Long-term survival after acute myocardial infarction is lower in more deprived neighbourhoods. Circulation 111:3063–3070

Vittinghoff E, Glidden DV, Shiboksi SC, McCulloch CE (2005) Regression methods in biostatistics: linear, logistic, survival, and repeated measures models. Springer, New York, Chapter 6.1, 6.2

Chapter 11
Modelling Time-to-Event Data

Kaplan-Meier Survival Analysis and Cox Regression

Gail M. Williams and Robert Ware

Abstract Much clinical research involves following up patients to an adverse outcome, which could be death, relapse, an adverse drug reaction or the development of a new disease. In these studies, time to event needs to be modelled such that factors that delay such events can be determined. The set of statistical procedures used to analyze such data is collectively termed survival analysis and is a very useful tool in clinical research. This chapter introduces the different tools of survival analysis.

Introduction

Survival analysis is concerned with describing and predicting the time to the occurrence of a binary event of interest, such as recovery or non-recovery from an illness, or development of a complication after surgery. This contrasts with logistic regression, which is concerned with the predictors of a binary event, irrespective of when it occurs. Survival analysis is typically more powerful than logistic regression, particularly when a high proportion of subjects experience the event, as it uses more information on the outcome.

In discussing these techniques, the time an event takes to occur is interchangeably referred to as survival time or failure time or time to event, and we refer to the event we are interested in as our outcome. This may be the achievement of any particular status: examples include death, death from a specific cause, relapse, recovery, or completion of a vaccination schedule. If a subject does not experience an event of interest in the follow-up period, the survival time is said to be censored.

G.M. Williams (✉) • R. Ware
School of Population Health, University of Queensland, Brisbane, QLD, Australia
e-mail: g.williams@sph.uq.edu.au

S.A.R. Doi and G.M. Williams (eds.), *Methods of Clinical Epidemiology*,
Springer Series on Epidemiology and Public Health,
DOI 10.1007/978-3-642-37131-8_11,

Example 1

The AIDS Clinical Trials Group carried out a randomized double-blind clinical trial on 1,151 patients with human immunodeficiency virus (HIV) infection. This trial compared the three-drug regimen of indinavir, zidovudine or stavudine, and lamivudine with the two-drug regimen of zidovudine or stavudine and lamivudine. Patients were eligible for the trial if they had no more than 200 CD4 cells/cm^3 and at least 3 months of prior zidovudine therapy. Because of the importance of CD4 cell counts in progression from HIV infection to AIDS, randomization was stratified by CD4 cell count at the time of screening. The primary outcome measure was time to a diagnosis of AIDS, or death from any cause.

Table 11.1 shows some relevant variables for the first 20 observations of data. The variable TIME is the survival time for the patient in days. The variable CENSOR identifies the outcome at the end of the follow-up period: if CENSOR = 1 the patient has died or been diagnosed with AIDS and the survival time is the follow-up time at the time of the event; if CENSOR = 0 then neither of these has occurred during the total follow-up time, and the survival time is the total duration of follow-up. Failure may still occur (at some time later than the follow-up time), but we do not know when. More specifically, this is sometimes referred to as right-censoring since the unknown values are at the right-hand end of the distribution of survival times. The variable TX records the treatment group: TX = 1 if the treatment included the third drug (indinavir) and TX = 0 if it did not. STRAT2 = 1 if CD4 cell counts were greater than 50/mm^3 and STRAT2 = 0 if not. So, for example, the patient with ID = 1 was followed for 189 days without progressing to a diagnosis of AIDS or death. Patient with ID = 14 progressed to a diagnosis of AIDS on day 206. Both patients who progressed to AIDS or died (ID = 14, 17) were in the low CD4 cell count group and did not receive the additional drug.

A Descriptive Analysis: The Kaplan–Meier Estimator

At first sight it might appear that we could simply estimate the mean survival time for the two treatment groups. The problem is that we only know this for those who experience the event, that is, 96 of the 1,151 patients in the trial. In the two-drug group, 63/577 = 10.9 % experienced an event and in the three-drug group it was 33/574 = 5.8 %; the mean survival times (time to AIDS or death) were 110 and 94 days, respectively. However, these times are an underestimate of survival time, because they are biased towards those who experience the event early; the longer survival times have yet to occur, and longer survival times will be more likely to occur in the more successful treatment group (if there is one).

It certainly appears that the three-drug treatment is more successful than the two-drug treatment, but we need a method that combines event occurrence and time. The solution to this problem lies in dividing the follow-up period into small time

Table 11.1 Example of survival data: AIDS trial

ID	TIME	CENSOR	TX	STRAT2	ID	TIME	CENSOR	TX	STRAT2
1	189	0	0	1	11	334	0	1	0
2	287	0	0	1	12	285	0	1	1
3	242	0	1	0	13	265	0	1	1
4	199	0	0	1	14	206	1	0	0
5	286	0	1	0	15	305	0	0	1
6	285	0	1	0	16	110	0	0	0
7	270	0	0	1	17	298	1	0	0
8	285	0	1	1	18	287	0	1	1
9	276	0	0	1	19	103	0	0	0
10	306	0	0	1	20	291	0	0	1

periods and comparing groups within each of these time periods. However, for now we examine the survival curve itself, using a technique that involves basic probability, to obtain an estimate of the probability of surviving to a specified time period. This is called the survivorship or just survival function.

We examine the 577 patients on the two-drug treatment. We set out the relevant data as in Table 11.2. As this table shows, we start at time $= 0$ with 577 patients, all yet to experience the event. For each time period (1 day), we count the number of patients whose follow-up ends (because they experience the event or are lost to follow-up at that point). For day 1, there are two such patients: one who was lost to follow-up (censored) and one who experienced the event (failed: diagnosis of AIDS or death). This leaves 575 who have data for times beyond day 1. Only one patient (who experienced the event) had a follow-up time of 2 days, leaving 574 patients; on day 3, one patient was lost to follow-up, and so on. This tabulation continues until the last time period – in this case day 364.

The reasoning used to get the survivorship function is as follows: we want the probability of surviving to a particular time period from a time $t = 1,2,\ldots$ to the end of the period. If we had complete data on everyone's survival, we could simply get the total number surviving to that day and divide by the number at the start. However, our at-risk denominator is changing due to censoring, so we use a simple probability model as follows:

The probability of surviving to time t

= the probability of surviving to time $t - 1$

× the probability of surviving the time period $t - 1$ to t

= the probability of surviving to time $t - 1$

× (1 – the probability of failing in the time period $t - 1$ to t)

The reason this works is that the second part on the right-hand side of the equation can be estimated easily from the number at risk of an event at time $t - 1$ and the number who failed from $t - 1$ to t and the first part on the right-hand side is simply the survival probability for the previous time $t - 1$.

Table 11.2 Aggregation of survival data to get the survival curve for 577 patients over 364 days of follow-up: first and last 10 days

Day	Number left at start of day	Censored	Failed	Day	Number left at start of day	Censored	Failed
1	577	1	1	.	.	.	.
2	575	0	1	356	2	0	0
3	574	1	0	357	2	0	0
4	573	1	0	358	2	0	0
5	572	0	0	359	2	0	0
6	572	0	0	360	2	0	0
7	572	1	1	361	2	0	0
8	570	0	0	362	2	1	0
9	570	0	1	363	1	0	0
10	569	1	0	364	1	1	0

At the start of day 1, no one has experienced an event so our survivor function is 1. We begin by working out the probability of surviving to the end of day 1. This is the probability of surviving to day 0 (1), multiplied by the probability of surviving to the end of day 1 (1 − 1/577 = 0.9983) = 0.9983. For the probability of surviving to the end of day 2, we have the probability of surviving to the end of day 1 (0.9983), multiplied by the probability of surviving day 2 (1 − 1/575 = 0.9983) = 0.9965. For the probability of surviving to the end of day 3, we have the probability of surviving to the end of day 2 (0.99653), multiplied by the probability of surviving day 3 (1) = 0.99653. We continue in this way to the end of the time period (Table 11.3). Note that the survival function only changes when at least one event occurs.

This process was first presented by Kaplan and Meier in 1958 and is referred to as the Kaplan–Meier or product–moment method for survival curves. The results can be plotted. Figure 11.1 shows how the first part of such a graph relates to the above data.

Circles indicate deaths and triangles indicate censored observations. The flat parts of the curve represent periods when no additional failures occur, and the down steps represent points at which failures occur. Note that censoring does not affect the occurrence of down steps but will affect the magnitude of the step because, if a large number of losses to follow-up occur, the following down step will be larger, as losses are removed from the at-risk denominator.

The formula for the standard error of the survival function at time t_i is

$$\hat{\sigma}(\hat{S}(t_i)) = \hat{S}(t_i)\sqrt{\sum_{j=1}^{i} \frac{d_j}{n_j s_j}}$$

where $\hat{S}(t_i)$ is the survival function at time i, n_j is the number at risk at time j, d_j is the number of failures at time j, and $s_j = n_j - d_j$.

Table 11.3 Survival function (estimated proportion surviving) for 577 patients over 364 days of follow-up: first and last 10 days

Day	Number left at start of day	Failed	Proportion failed	Survival function	Day	Number left at start of day	Failed	Proportion failed	Survival function
0	577			1	.	.	.	.	.
1	577	1	0.00173	0.9983	355	2	0	0	0.8692
2	575	1	0.00347	0.9965	356	2	0	0	0.8692
3	574	0	0	0.9965	357	2	0	0	0.8692
4	573	0	0	0.9965	358	2	0	0	0.8692
5	572	0	0	0.9965	359	2	0	0	0.8692
6	572	0	0	0.9965	360	2	0	0	0.8692
7	572	1	0.00695	0.9930	361	2	0	0	0.8692
8	570	0	0	0.9930	362	2	0	0	0.8692
9	570	1	0.00870	0.9913	363	1	0	0	0.8692
10	569	0	0	0.9913	364	1	0	0	0.8692

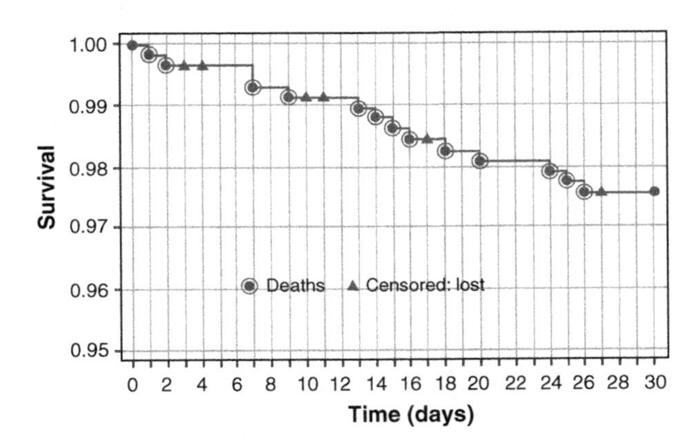

Fig. 11.1 Survival function (estimated proportion surviving), for 577 patients over the first 30 days

This can be used to obtain confidence limits for the survival function. Figure 11.2 shows the complete survival curve, with 95 % confidence limits for the estimated survival at each time point. This particular survival curve shows a more or less steady decline in survival over time.

It is sometimes useful to look at the cumulative incidence of events; this is just the inverse of the survival function, and can be calculated using the formula:

$$\text{Cumulative Incidence} = 1 - \text{Survival}$$

Figure 11.3 shows the estimated cumulative proportion of persons experiencing the event over time. Notice the vertical axis has been reduced in scale, allowing more detail to be shown, compared with the survival curve, which typically uses a vertical axis from 0 to 1.

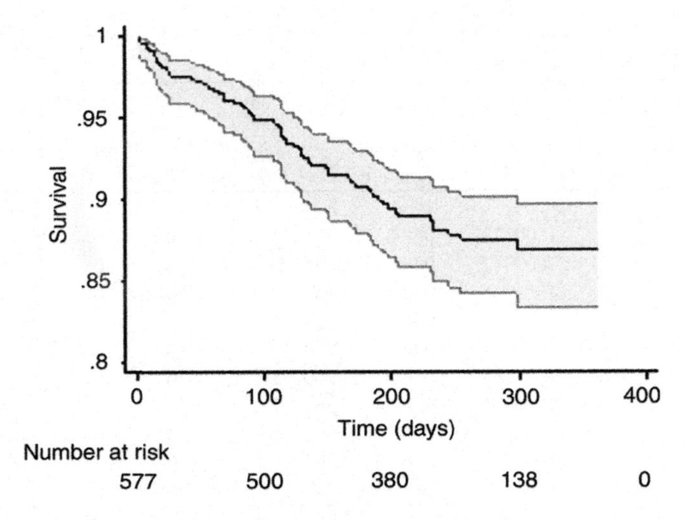

Fig. 11.2 Survival curve: time to AIDS diagnosis or death for two-drug treatment group, with 95 % confidence limits

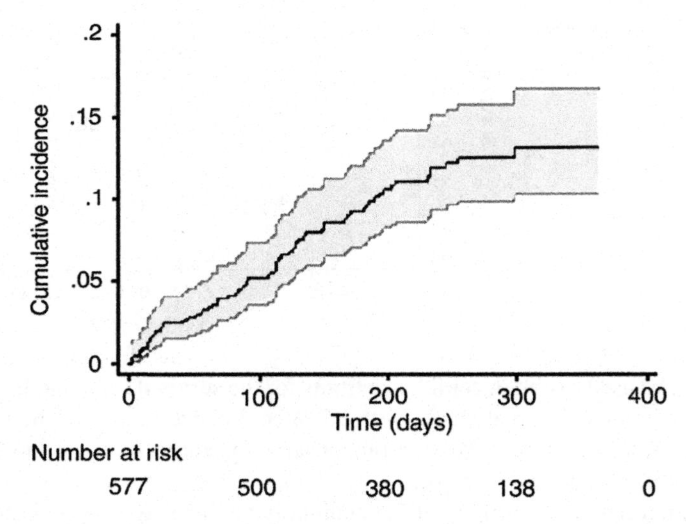

Fig. 11.3 Cumulative incidence of AIDS diagnosis or death for two-drug treatment group, after recruitment, with 95 % confidence limits

Let us now return to our primary research question: is there a difference between the two drug treatments in terms of patient progression to AIDS, or death? We need to compare the two treatment groups. We first plot the survival curve for each group, with 95 % confidence limits.

Figure 11.4 shows that the survival (alive or free of a diagnosis of AIDS) is better for the three-drug treatment. By the end of the period of observation, the survival

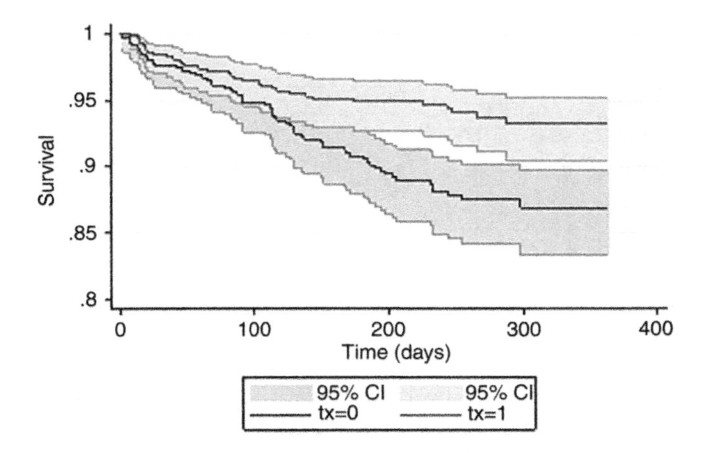

Fig. 11.4 Survival curves: time to AIDS diagnosis or death for the two-drug treatment group (tx = 0), compared with the three-drug treatment group (tx = 1) with 95 % confidence limits

for the two-drug treatment group is around 87 %; for the three-drug treatment it is around 93 %. The 95 % confidence limits also begin to separate at about 6 months after treatment. Figure 11.5 shows the relative cumulative incidence of AIDS diagnosis or death for the three-drug treatment group, compared with the two-drug treatment group. This is simply the ratio of the cumulative incidence in the three-drug group to that in the two-drug group. We see that this is quite variable up to about 2 months (due to the small number of events) and then the three-drug treatment becomes progressively more protective until the relative efficacy stabilizes at about 50 % from about 6 months.

We can also proceed to a statistical test to examine formally whether this difference in survival is due to chance.

Log-Rank Test for Comparisons of Groups

The formal test to determine whether survival is different for two treatment groups is the log-rank test. The null hypothesis for this test is that the distributions of survival times within the two treatment groups are the same, the alternative hypothesis being that they differ in some way. This test is a non-parametric test; it makes no assumptions about any particular distribution of survival times. It involves calculating the expected numbers of events (say deaths) in each group, under the assumption that the two survival distributions are the same. It does this by calculating the total number of deaths at each time period, then splitting this number of deaths at each time period according to the numbers at risk in each comparison group, thereby forming expected numbers of deaths, under the assumption the null hypothesis is true. For example, if there were four deaths on day 100, with group

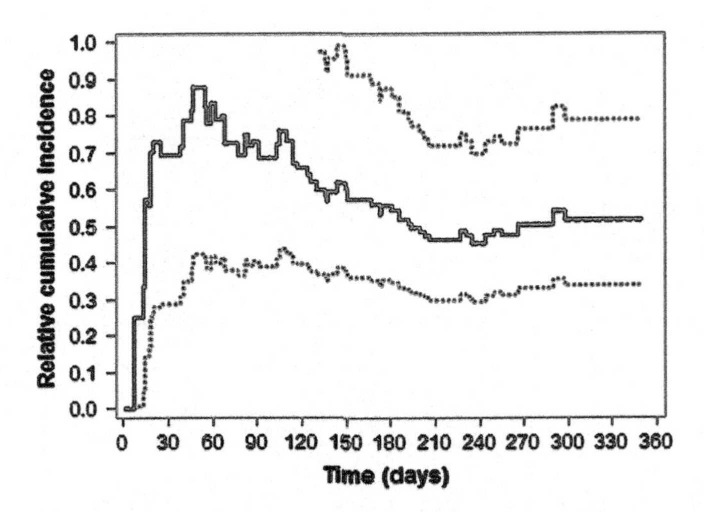

Fig. 11.5 Relative cumulative incidence of AIDS diagnosis or death for the three-drug treatment group, compared with the two-drug treatment group, with 95 % confidence limits (lighter plots)

1 having 500 people still at risk (after deaths and dropouts before day 100) and group 2 having 300 people still at risk, then the expected deaths would be 2.5 (= 4 × 500/800) in group 1 and 1.5 in group 2. (Note that fractional expected events are permitted). All expected deaths are then added up within each group to get the overall expected deaths for each group. These are compared with the observed number of deaths. If the numbers of expected and observed deaths are similar, this supports the null hypothesis. If they are quite different, this would be regarded as evidence against the null hypothesis.

Table 11.4 shows these calculations for the first 30 days of the AIDS trial: for example, the three deaths on day 14 (1 in the two-drug group and 2 in the three-drug group) are split according to the ratio of persons in the study on those days: 566:570 or 1.49–1.51. A chi-squared-like formula based on the differences between observed and expected events in the two groups is then used to get the test statistic, which is then referred to a chi-squared distribution. For the AIDS study, this yields chi-squared = 10.54, with 1 df, giving $P = 0.0012$, indicating a statistically significant difference in survival between the groups.

Quantification of the difference in survival is a little problematic, because the Kaplan–Meier method is a non-parametric method and does not involve any natural epidemiological measures of effect, such as relative risks. However, we can note that the survival curves demonstrate 6-month survival of 90.4 % (95 % CI 87.8–94.0 %) in the two-drug group and 94.9 % (95 % CI 93.0–96.8 %) in the three-drug group, with 12-month survivals of 86.9 % (95 % CI 83.7–90.1 %) and 93.2 % (95 % CI 90.8–95.6 %), respectively. This amounts to an approximate doubling of the cumulative incidence at each of these time points. In general, however, such comparisons may result in different relative cumulative incidence at each time period.

Table 11.4 Calculations for expected events

Time	Observed deaths: two drugs	Observed deaths: three drugs	Total deaths	At risk: two drugs	Expected deaths: two drugs	At risk: three drugs	Expected deaths: three drugs
1	1	0	1	577	0.5013032	574	0.4986968
2	1	0	1	575	0.5008711	573	0.4991289
7	2	1	3	572	1.5000000	572	1.5000000
9	1	0	1	570	0.4995618	571	0.5004382
13	1	1	2	567	0.9964851	571	1.0035149
14	1	2	3	566	1.4947183	570	1.5052817
15	1	0	1	565	0.4986761	568	0.5013239
16	1	0	1	564	0.4982332	568	0.5017668
17	0	1	1	563	0.4977896	568	0.5022104
18	1	2	3	562	1.4946809	566	1.5053191
20	1	1	2	561	0.9982206	563	1.0017794
24	1	0	1	561	0.4995548	562	0.5004452
25	1	1	2	559	0.9973238	562	1.0026762
26	1	0	1	557	0.4982111	561	0.5017889
Total over 364 days	63	33	96		47.1		48.9

The log-rank test generalizes to more than two groups, yielding chi-squared tests with df = the number of groups being compared − 1.

Example 2

As a second example, Fig. 11.6 shows the survival curve for a series of 500 patients from the Worcester Heart Attack Study, which examined factors associated with survival rates following hospital admission for acute myocardial infarction (MI). Data were collected during 13 one-year periods beginning in 1975 and extending until 2001, on all MI patients admitted to hospitals in the Worcester, Massachusetts Standard Metropolitan Statistical Area.

Of the 500 patients, 215 (43 %) died within their follow-up period, with a median survival time of 5.5 months, and 285 were censored (still alive at the end of their follow-up period), with a median follow-up time of 3.4 years. Notice that mortality is highest immediately after the initial MI, with only about 70 % surviving the first year, but flattens out a little at about 2 years. This analysis yields more information than simply examining mortality rates. This survival curve can also be used to estimate the median survival, which is the length of time corresponding to 50 % survival: draw a horizontal line at survival = 0.5, then project this vertically to the horizontal axis to give an estimate of years; thus, an estimated 50 % of persons survive to 5 years or more (or the 5-year survival is 50 %) (Fig. 11.6). Other quantiles can be estimated similarly. For example, the 75 % survival is about

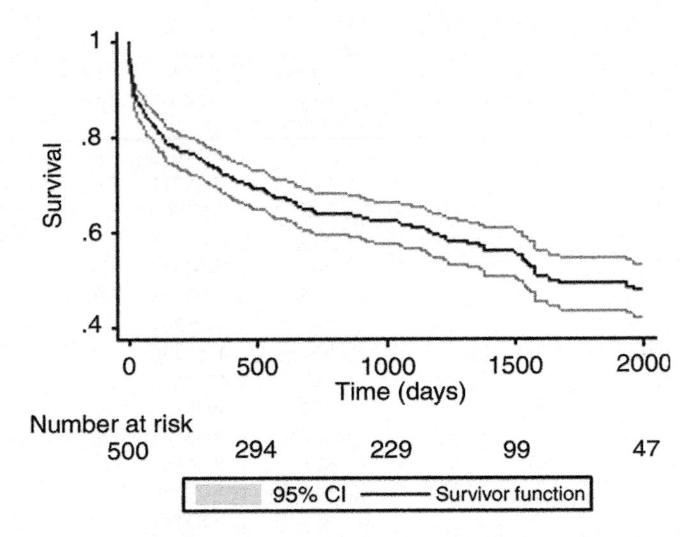

Fig. 11.6 Survival curve: time to death following an MI for 500 patients in the Worcester Heart Attack Study, with 95 % confidence limits

6 months and the 25 % survival is 6.5 years. There may be a dramatic drop at the end of the series (corresponds to the death of the one person remaining in the study, all others having died or been lost to follow-up). This is not seen in Fig. 11.6 because time on the X-axis has been truncated at 2000 days.

Additional Comments on Censoring

The above examples involve censoring as a consequence of loss to follow-up. However, it is possible to have censoring for other reasons. Essentially, whenever a participant experiences an event that precludes them from experiencing the event of interest, they are censored at that point. For example, if in the Worcester Heart Attack Study our event of interest is death from cardiovascular disease, then a patient who experiences death from cancer is considered censored at the time of his/her death.

We also assume that censoring is independent; that is, observations that are censored have the same future risk as those that are not censored. This is quite a strong assumption. For example, suppose we are examining survival of patients admitted to an intensive care unit. If follow-up ceases at discharge then, given that those discharged are more likely to have a better prognosis than those not, censored observations would be likely to have a lower risk. If dropouts differ from non-dropouts, then this is also likely to lead to dependent censoring. In the Worcester Heart Attack Study, it may be that patients experiencing complications following MI would be more likely to be in contact with a medical practitioner and thus less likely to be lost to follow-up.

A specific example of this occurs in 'competing risk' models. In the example given above for death from cardiovascular disease, and censoring by death from cancer, we would need to consider whether a person who dies from cancer would have the same future risk (had s/he not died from cancer) of death from cardiovascular disease. This may seem plausible. However if, in a series of patients with obesity, the event of interest is death from cardiovascular disease, then death from diabetes would be a competing event, since it is likely that risks of these outcomes would be associated. In such instances it is argued that the methods described above would result in biased estimates of survival. The appropriate analyses are complex, essentially involving the joint analysis of the cumulative incidence of both the event of interest and any competing event(s).

Multivariable Survival Analysis

The Kaplan–Meier method allows us to examine and compare survival across groups. It does so by stratification into groups and using the log-rank test to make formal statistical comparisons. This approach is very flexible but not suitable for multivariable situations when we have a large number of variables, particularly confounders, for which we need to adjust. A modelling approach that combines traditional regression methods with the technique applied for the Kaplan–Meier estimates provides an effective strategy. The follow-up time is divided into small periods, and then effects (relative risks of the event) are averaged over these time periods, within which a model is applied. An important assumption underlies this approach – that of proportional hazards – and modifications may be applied when this assumption is questionable.

Example 2 (Continued)

We return to the data from the Worcester Heart Attack Study, which recruited patients at the time of a first MI. The primary outcome measure for this trial was time to death. We would aim to identify factors that predict survival after MI. One of these might be the absence of complications at the time of the MI. In particular, we look at the presence or absence of congestive heart failure (CHF) complications.

Of the 500 persons entered into the study, 155 (31 %) had CHF complications; of the 215 deaths, 110 (51 %) were in the CHF complications group, indicating higher mortality in the CHF group. Of the 285 censored events (follow-up finished before death occurred), 45 (29 %) were in the CHF group, indicating an equal rate of censoring for both groups.

The Kaplan–Meier survival curves are shown in Fig. 11.7, with 95 % confidence limits. It is clear that those without complications experience far better survival following their MI. Moreover, it appears that this may vary somewhat over time;

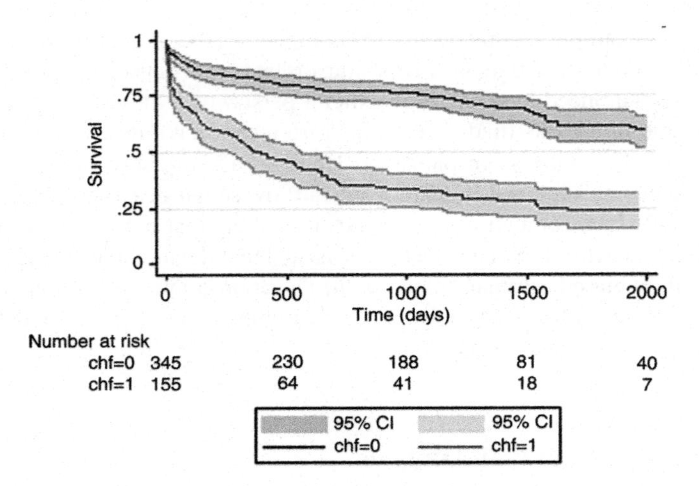

Fig. 11.7 Kaplan–Meier survival curves: time to death, comparing those with (chf = 1) and without (chf = 0) CHF complications, with 95 % confidence limits

those with complications appear to have even greater mortality soon (up to around 2 years) after the MI, after which the survival curves are approximately parallel.

The log-rank test confirms the higher mortality in the CHF complications group, with a chi-squared of 84.6, 1 df, and $P < 0.0001$.

We now introduce alternative approaches to examining such data, so that we can (a) quantify difference in risks and (b) examine multiple risk factors simultaneously, allowing for adjustments by confounding factors or other risk factors.

Hazard Function and Hazard Ratio

To proceed with a modelling approach, we require the concept of a hazard function, defined as the short-term event rate for participants who have not yet experienced the event of interest. The hazard function, for example, describes the probability of death on a particular day in our case study of the Worcester Heart Attack Study. We saw that the hazard function depended on time; it appeared that the risk of death was higher in the immediate post-MI period and then decreased thereafter (Fig. 11.6). The advantage of now defining this more generally is that we can extend our framework to include a hazard function that depends on a number of variables, such as risk factors or confounders, setting the scene for multivariate analysis. Note the difference between the hazard function and cumulative failure. The hazard function represents risk at a point in time, the cumulative failure represents risk up to a point in time.

Let us consider patients who experienced CHF complications, compared with those who did not. Figure 11.7 shows that the mortality was considerably lower for patients without complications, compared with those with complications, and that the gap between these two groups arose quite quickly after the initial heart attack. After about 2 years, the survival courses become approximately parallel.

We now define the hazard ratio as the ratio of these two hazard functions, as follows.

$$\text{Hazard ratio} = \frac{\text{Hazard function for patients with complications at time } t}{\text{Hazard function for patients without complications at time } t}$$

Cox Proportional Hazards Model

The Cox proportional hazards model is a widely used method for multivariate analysis of survival data. It can be thought of as a combination of the Kaplan–Meier method, which involves dividing the follow-up time into small intervals, and the regression model approach, which formulates a model within each of these time periods.

Let us suppose our follow-up time is divided into intervals determined by the time points t_1, t_2, t_3, etc. We imagine a regression model within each of these intervals (t_1, t_2) etc., relating the probability of death within the interval to a risk factor, such as CHF complications. From such a model we can obtain an effect estimate (a relative risk of death in this case, associated with having CHF complications) for that time interval. In general, this relative risk varies over time; we have seen a suggestion of this in the CHF complications example. However, we begin by making a simplifying assumption, for the time being, that the hazard ratio is constant over time. For now, we can think of this constant hazard ratio as the average hazard ratio over time, and we explore variability later. Note that we can allow the hazard functions to vary over time; we only require that the hazard ratio is a constant (i.e. does not depend on time). This assumption is called the proportional hazards assumption.

We now consider a particular model, to cover the time period from t to $t + 1$, as follows.

$$\begin{aligned} &\log(\text{Number of deaths at time } t) \\ &\quad = a(t) + b \times (\text{CHF} = \text{Present}) + \log(\text{PTAR within time } t) \end{aligned}$$

The term $a(t)$ corresponds to the usual intercept term (depends on the time t) and the term b is the usual slope estimate attached to the predictor variable, which in this case is presence/absence of complications. The term (CHF = Present) takes the value 1 when the statement within parentheses is true and is zero otherwise. It is thus a dummy variable for the presence of complications. The PTAR is the person-time at risk within the time period t. If we have 1 – day time periods and $t = 100$,

Table 11.5 Cox regression for mortality after an MI by the presence or absence of CHF complications

Parameter		Parameter estimate	P value	Hazard ratio (HR)	95 % CI for HR	
CHF complications		Reference category: Absent				
Present	b	1.198	<0.0001	3.31	2.53	4.34

then PTAR in days = the number of persons still in the study at the start of the 100th day. The terms involving a and b are the usual intercept and regression coefficients used in previous regression models. We note the log transform on the left-hand side. If we reorganize a little we have the following:

$$\log\left(\frac{\text{Number of deaths at time } t}{\text{PTAR within time } t}\right) = \log(\text{Death rate at time } t)$$
$$= a(t) + b \times (\text{CHF} = \text{Present})$$

Conceptually, this model is then fitted for each time period (with the constraint that b is the same for each period). This occurs by a process called partial likelihood. The resultant parameter b is then the log of the hazard ratio, defined above.

The usual form of the model expresses the intercept as a log of a baseline hazard, this being the predicted death rate for the reference category of the predictor variable.

$$\log(\text{Death rate at time } t) = \log D_0(t) + b \times (\text{CHF} = \text{Present})$$

In general the baseline hazard is the predicted death rate at a particular time when an individual is in the reference category for all predictor variables. Thus, the hazard ratio gives the multiplier of the baseline hazard associated with being in a particular risk group. You can see that the hazard ratios multiply, because the above model, after taking the exponential of both sides, becomes

$$\text{Death rate at time } t = D_0(t) \times \exp(b \times (\text{CHF} = \text{Present}))$$

When we fit the model, we find the following (Table 11.5). The parameter estimate is the estimate of b in the above model, and the hazard ratio is obtained by taking the exponential of the parameter estimate.

The hazard ratio is 3.31 with 95 % confidence limits of 2.53–4.34. It should be emphasized that the hazard ratio tells us nothing about how fast a person dies. Accordingly, a hazard ratio of 3.31 should not be interpreted as showing that patients in the CHF group died about three times as fast as those without CHF. In survival terms, it does not imply that the median survival time was cut to a third by the treatment nor that three times as many patients were likely to have died on or by a particular day or that the CHF group was likely to have died three times as rapidly as those without CHF. The correct interpretation is that for any randomly selected

Table 11.6 Cox regression for mortality after an MI by patient's sex

Parameter		Parameter estimate	P value	Hazard ratio (HR)	95 % CI for HR	
Patient sex	b	Reference category: male				
Female		0.381	0.0056	1.46	1.12	1.92

pair of patients, one from the CHF group and one from the group without CHF who have not died by a certain time, the hazard ratio of 3.31 represents the chance of death in the patient from the CHF group at the next point in time compared to the patient from the group without CHF.

We may consider it likely that CHF complications might be distributed differently over age at MI and sex, and these factors may also affect survival. Let us examine the effect of the patient's sex by fitting a Cox model. This yields Table 11.6.

Thus females have a higher risk of death after MI, on average, than males.

Age could be examined in two ways: as a continuous variable (assuming a linear effect on the log-risk scale) or in groups. The models for these are written as follows:

$$\log(\text{Deathrateat time } t) = \log D_0(t) + b \times \text{Age}$$

$$\begin{aligned}\log(\text{Death rate at time } t) = \log D_0(t) &+ b_1 \times (\text{Age} = \text{65-79 years}) \\ &+ b_2 \times (\text{Age 80 years or greater})\end{aligned}$$

See Table 11.7 for results of the Cox regression for each of these.

Thus there is a fourfold increase in risk of mortality for persons aged 65–79 years and an eightfold increase in risk of mortality for persons 80 years of age or older, compared with those less than 65 years. These increases are all clearly significant, with 95 % confidence limits that exclude 1.

Since presence of CHF may be associated with age and sex, which themselves relate to survival, we now wish to consider a multivariable model which examines the effect of CHF, after adjusting for age and sex. This model may be written as follows.

$$\begin{aligned}\log(\text{Death rate at time } t) = \log D_0(t) &+ b \times (\text{CHF} = \text{Present}) \\ &+ c \times \text{Age} \\ &+ d \times (\text{Sex} = \text{Female})\end{aligned}$$

We see that the age effect remains much as before – a slightly increased (6 %) risk of mortality per year of age. The sex effect, however, has reduced, so that females now have a lower mortality than males, although this is not significant. The adverse effect of CHF complications remains, although reduced to a hazard ratio of 2.4 from 3.3 (Table 11.8).

A comparison of males and females shows the following (Table 11.9).

Table 11.7 Cox regression for mortality after a myocardial infarct by patient age

Parameter		Parameter estimate	P value	Hazard ratio (HR)	95 % CI for HR	
Patient age						
Age in years	b	0.066	<0.0001	1.068	1.056	1.081
Age group		Reference category: Age < 65 years				
65–79 years	b_1	1.411	<0.0001	4.10	2.57	6.53
80+ years	b_2	2.078	<0.0001	7.99	5.09	12.55

Table 11.8 Cox regression for mortality after an MI by patient age at MI, sex and presence/absence of CHF complications

Parameter		Parameter estimate	P value	Hazard ratio (HR)	95 % CI for HR	
CHF complications		Reference category: Absent				
Present	b	0.878	<0.0001	2.41	1.82	3.19
Patient age (years)	c	0.060	<0.0001	1.062	1.049	1.075
Patient sex		Reference category: Male				
Female	d	−0.155	0.273	0.86	0.65	1.13

Table 11.9 Characteristics of males and females

	Males	Females
N	300	200
Mean age (range) in years	66.6 (30–102)	74.7 (32–104)
CHF complications, n (%)	75 (25)	80 (40)

Thus females were older than males and more likely to have CHF. Both of these characteristics increase the risk of mortality. When adjusted for these additional risks, males and females had similar post-MI survival.

Testing the Proportional Hazards Assumption

Recall the proportional hazard assumption, which assumes a constant rate (here 3.31). We might conclude from our earlier Kaplan–Meier and Life Table analyses that the proportional hazards assumption is likely to be violated in the case of CHF, and that a better-fitting model would take this into account and allow a different hazard ratio in the first 2 years and the remaining 2years. We can examine this formally within the Cox regression approach.

We do this by adding a time-dependent term to our model. A time-dependent term is one that varies over time: CHF complications at the start of the study is a non-time-dependent factor; it remains constant. However, time since follow-up obviously varies with time; other factors that vary with time might be age or subsequent complicating conditions. Note that some variables might be treated as either; for example, age at baseline is not time dependent, age at follow-up time is. How we use a particular variable depends on how we frame our research question.

The particular term we add is an interaction term: we first define a time-dependent variable as follows:

$$\text{Period} = 0 \ \text{ if follow-up time at time } t \text{ is less than 2.5 years}$$
$$\text{Period} = 1 \ \text{ if follow-up time at time } t \text{ is 2.5 years or more}$$

We then add an interaction term Period × CHF, which has the effect of allowing the effect of CHF to vary between the two periods defined above. A model is now written as follows.

$$\log(\text{Death rate at time } t) = \log D_0(t) + b \times (\text{CHF} = \text{Present}) + c \times (\text{CHF} = \text{Present}) \times (\text{Period (t)} = 1)$$

Note that the model formulation makes explicit that the value of Period depends on the follow-up time. Recall that a proportional hazards model involves splitting the follow-up time into small intervals of time, and fitting a model within each of those times. Thus the value of Period will be zero for all intervals falling within the follow-up period up to 2.5 years and 1 subsequently. The b parameter represents the effect of CHF complications on mortality within the first 2.5 years, and the c parameter represents the difference between the CHF effects for the two time periods. This can be seen by noting that the model becomes

$$\log(\text{Death rate at time } t) = \log D_0(t) + b \times (\text{CHF} = \text{Present})$$

for any time before 2.5 years, and

$$\log(\text{Death rate at time } t) = \log D_0(t) + (b + c) \times (\text{CHF} = \text{Present})$$

for any time after 2.5 years.

The results of fitting this model are shown in Table 11.10.

These results must be interpreted carefully. Remember that interactions are included in models to allow the effects of variables to differ. We include the CHF × Period interaction because we want to examine the possibility of different CHF effects in the two time periods. Thus, by definition, there is no single parameter that measures the CHF effect. The first hazard ratio (3.78) given in Table 11.10 is that for the first period, up to 2.5 years. The second hazard ratio (0.45) is the ratio of the hazard ratio in the second period, compared with that in the first. So the hazard ratio for the second period is 0.45 × 3.78 = 1.70. The P value of 0.044 for the interaction term informs us that the two hazard ratios (for the first and second period) are statistically significantly different. We then obtain 95 % confidence limits for the estimate of 1.70 in period 2 as (0.83–3.48), indicating it is not statistically significant. We can express our final results as in Table 11.11.

Table 11.10 Cox regression for mortality after an MI by the presence or absence of CHF complications, allowing a time-dependent interaction between CHF complications and follow-up time up to less or greater than 2.5 years (Period)

Parameter		Parameter estimate	*P* value	Hazard ratio	95 % CI for HR	
CHF complications		Reference category: Absent				
Up to 2.5 years: Present	*b*	1.328	<0.0001	3.78	2.80	5.09
CHF × Period	*c*	−0.800	0.044	0.45	0.21	0.98

Table 11.11 Cox regression for mortality after an MI by the presence or absence of CHF complications before and after 2.5 years

Parameter		Parameter estimate	*P* value	Hazard ratio	95 % CI for HR	
CHF complications		Reference category: Absent				
Present: up to 2.5 years	*b*	1.328	<0.0001	3.78	2.80	5.09
Present: after 2.5 years	*b* + *c*	0.528	0.150	1.70	0.83	3.48

Sometimes a lack of proportional hazards may be caused by omission of an important covariable in the analysis. Age at MI is certainly an important covariable in relation to mortality and age is related to risk of CHF complications: the percentages of CHF complications in age groups under 65, 65–79, and over 80 years were 14.5 %, 34.3 % and 45.1 %, respectively. So we might consider whether a model which includes age would better satisfy the proportional hazards assumption. This model is written as

$$\begin{aligned}\log(\text{Death rate at time } t) = \log D_0(t) &+ b \times (\text{CHF} = \text{Present}) \\ &+ c \times (\text{CHF} = \text{Present}) \times (\text{Period (t)} = 1) \\ &+ d \times \text{Age}\end{aligned}$$

Table 11.12 shows the result of a model including age and the time-dependent interaction between CHF and follow-up time less than or greater than 2.5 years, and Table 11.13 shows the stratified presentation of this model. Although we still see a lower hazard ratio in the period after 2.5 years, this difference is no longer significant ($P = 0.178$, Table 11.12). Note that the inclusion of age in the model has reduced the CHF effect (in both time periods, but more noticeably in the first time period), as we would expect, given the association of age with both mortality and the presence of CHF complications.

Conclusion

The Kaplan–Meier method is a flexible informative technique for estimating survival over time, particularly when data are censored. Care must be taken in considering whether censoring is dependent on factors that may affect survival.

Table 11.12 Cox regression for mortality after an MI by the presence or absence of CHF complications, allowing a time-dependent interaction between CHF complications and follow-up time up to less or greater than 2.5 years (Period), and including age at MI

Parameter		Parameter estimate	*P* value	Hazard ratio (HR)	95 % CI for HR	
CHF complications		Reference category: Absent				
Present: up to 2.5 years	*b*	0.949	<0.0001	2.58	1.90	3.52
CHF × Period	*c*	−0.539	0.178	0.583	0.27	1.28
Patient age (years)	*d*	0.058	<0.0001	1.060	1.047	1.072

Table 11.13 Cox regression for mortality after an MI by the presence or absence of CHF complications before and after 2.5 years, adjusting for age at MI

Parameter		Parameter estimate	P value	Hazard ratio (HR)	95 % CI for HR	
CHF complications		Reference category: Absent				
Present: up to 2.5 years	*b*	0.949	<0.0001	2.58	1.90	3.52
Present: after 2.5 years	*b* + *c*	0.410	0.267	1.51	0.73	3.11
Patient age (years)	*d*	0.058	<0.0001	1.060	1.047	1.072

A regression-based extension of the Kaplan–Meier method is the Cox proportional hazards model, which can be used to assess the effect of multiple covariates on survival. It is a very flexible method for handling clinical data, especially for prognostic clinical research in which the influence of multiple variables on survival is to be evaluated.

Bibliography

Cox DR (1972) Regression models and life tables. J R Stat Soc 34:187–220

Hammer SM, Squires KE, Hughes MD, Grimes JM, Demeter LM, Currier JS, Eron JJ Jr, Feinberg JE, Balfour HH Jr, Deyton LR et al (1997) A controlled trial of two nucleoside analogues plus indinavir in persons with human immunodeficiency virus infection and CD4 cell counts of 200 per cubic millimeter or less. AIDS Clinical Trials Group 320 Study Team. N Engl J Med 337:725–733

Kalbfleisch JD, Prentice RL (1980) The statistical analysis of failure time data. Wiley, New York

Kaplan EL, Meier P (1958) Nonparametric estimation from incomplete observations. J Am Stat Assoc 53:457–481

Spruance SL, Reid JE, Grace M, Samore M (2004) Hazard ratio in clinical trials. Antimicrob Agents Chemother 48:2787–2792

Tonne C, Schwartz J, Mittleman M, Melly S, Suh H, Goldberg R (2005) Long-term survival after acute myocardial infarction is lower in more deprived neighbourhoods. Circulation 111:3063–3070

Part IV
Systematic Reviews and Meta-analysis

Chapter 12
Systematic Reviewing

Introduction, Locating Studies and Data Abstraction

Justin Clark

Abstract A systematic review is essentially a systematic investigation of existing research data identified via a reproducible systematic search leading to data abstraction, appraisal of methodological quality, clinical relevance and consistency of published evidence on a specific clinical topic in order to provide clear suggestions for a specific health care problem. This can be followed by a quantitative synthesis, which, preserving the identity of individual studies, tries to provide an estimate of the overall effect of an intervention, exposure or diagnostic strategy. The latter is called a meta-analysis. This chapter outlines the procedure that needs to be followed to execute a standard systematic review.

Introduction

The systematic review process begins with the definition of a question and a hypothetical solution. There is then a problem formulation (population, intervention or exposure, comparison, outcome [PICO]), data search, data abstraction and appraisal, data analysis ± quantitative synthesis, and finally result interpretation and dissemination. After definition of the question according to the PICO approach, the appropriate keywords are used to search several databases. Useful resources include PubMed Central, Clinicaltrials.gov, EMBASE, LILACS, PubMed, conference proceedings, cross-referencing (hand searching) and contact with experts. It is critical to have explicit inclusion and exclusion criteria: The broader the research domain, the more detailed they tend to become and it is good to refine criteria as you

J. Clark (✉)
University of Queensland, Brisbane, Australia
e-mail: j.clark4@Uq.edu.au

S.A.R. Doi and G.M. Williams (eds.), *Methods of Clinical Epidemiology*,
Springer Series on Epidemiology and Public Health,
DOI 10.1007/978-3-642-37131-8_12,

interact with the literature. The components of detailed criteria are distinguishing features, research respondents, key variables, research methods, cultural and linguistic range, time frame and publication types. We do not only include peer-reviewed studies because significant findings are more likely to be published than non-significant findings; therefore, look for unpublished non-significant studies, e.g. abstracts in conferences. It is critical to try to identify and retrieve all studies that meet your eligibility criteria. Potential sources for identification of documents include computerized bibliographic databases, authors working in the research domain, conference programs, dissertations, review articles, hand searching relevant journal articles, government reports, bibliographies and clearing houses. In summary, the systematic review process requires definition of a focused question, a pre-specified protocol with strict inclusion and exclusion criteria, selection of the right type of studies, a comprehensive information-finding strategy, a detailed and objective data abstraction form, quality appraisal of all included studies and finally synthesizing and summarizing the information with a meta-analysis (if appropriate) followed by interpretation.

The systematic review imposes a discipline on the process of summing up research findings, represents findings in a more differentiated and sophisticated manner than conventional reviews, is capable of finding relationships across studies that are obscured in other approaches, protects against over-interpreting differences across studies and most important, can handle a large numbers of studies (this would overwhelm traditional approaches to review). Unfortunately, it also requires a good deal of effort; mechanical aspects do not lend themselves to capturing more qualitative distinctions between studies. There is the "apples and oranges" criticism and most systematic reviews include blemished studies to one degree or another (e.g. a randomized design with attrition). Also, selection bias poses a continual threat. Studies for which there were negative or null findings that were not reported are also a threat.

Locating Studies: A Step by Step Guide

This section takes you through the steps of constructing a clinical question and then makes use of that question to guide your search of the medical literature via the various databases available. Some key components of this data extraction process are examined via a step by step approach using a systematic review co-written by the author as an example.

When undertaking a systematic review, there are important things to be taken into consideration when searching for studies. These include:

1. Searching broadly
2. Ensuring your results can be reproduced

Searching broadly means that you should try to capture as much information on your topic as possible. Without a broad search of the literature you may miss potential information that you need in order to arrive at as accurate a result as possible.

Ensuring that your results are reproducible by others is a staple of research. Whether it is an experiment in a laboratory, gathering information from the field or, as in this case, finding and amalgamating information from other sources, it is detrimental to the believability and credibility of your research if it is not possible for someone else to come along, follow the same steps that you took and not get the same result. This section provides a step by step guide on how to do exactly that. It covers how best to layout your search, where to search and what is needed to ensure others can check your findings.

The Clinical Question

The first step in a systematic review is to construct a clinical question. This is a succinct statement laying out exactly what question you hope to answer with your systematic review. For example:

1. Does vitamin C help to prevent the common cold?
2. Does Tamiflu help to prevent the flu?
3. Does St Johns Wort help to prevent depression?

A clinical question should cover the following four areas, the type of person involved (normally associated with the health problem they are faced with), the type of exposure that the person has experienced (which could be a risk factor, prognostic factor, intervention, or diagnostic test), the type of control or comparison with which the exposure is being compared and the outcomes to be addressed. It is common practice, especially for those new to the systematic review process, to write it out using the following format:

- (P) the types of population or participants
- (I) the types of interventions
- (C) the types of comparisons
- (O) the types of outcomes

This is known as the PICO format and is used as a guide to ensure you have covered the topics required to answer your question. For example, the PICO format for our above questions is:

- P: People with a cold
- I: Vitamin C
- C: No vitamin C
- O: Reduction in cold symptoms and/or duration

- P: People with the flu
- I: Tamiflu
- C: No Tamiflu

- O: Reduction in flu symptoms and/or duration

- P: People with depression
- I: St Johns Wort
- C: No St Johns Wort
- O: Reduction in depressive episodes

The above are purposefully simplistic to provide an easy to follow example of the clinical question and the PICO format. In this chapter, we use a more complicated clinical problem to illustrate how to search the literature and locate studies.

The problem is this. Radioiodine (^{131}I) is widely used for the diagnosis and treatment of benign thyroid diseases. Observational studies have not been conclusive about the carcinogenic potential of ^{131}I and therefore we wish to undertake a systematic review and meta-analysis to determine if we can answer this question. To begin, we lay out the problem in the PICO format.

- P: Patients with benign thyroid disease
- I: Exposure to radioactive iodine in a medical setting, i.e. at a hospital
- C: No exposure to radioactive iodine
- O: Increase in cancer incidence

From this we can construct the following clinical question: Does medical exposure to radioactive iodine in benign thyroid disease increase the risk of cancer?

We started with a problem; we broke this problem up into some individual components using the PICO method to ensure that we covered the required information and then constructed an answerable question we can now use to guide the next steps of our systematic review and meta-analysis.

Constructing the Search

It is now time to let our question guide our search of the medical literature. You cannot just type the entire search string into the database and use whatever comes up. You need to break down the clinical question into individual search components capable of being inserted into the databases. These can be individual words or a combination of two or three words.

We begin by consulting our clinical question: Does medical exposure to radioactive iodine in benign thyroid disease increase the risk of cancer? From our question, we identify the key components, words or concepts that we are going to use to create our search. For this process, the PICO table you may have created can be used to help. We see that cancer is an important element that we need to search for. Exposure to radioactive iodine is also important. Benign thyroid disease is also a key component that we need to ensure is part of the search criteria. This gives us three key search criteria for use in the database:

1. Cancer
2. Radioactive iodine
3. Benign thyroid disease

You may have noticed that we had "no exposure to radioactive iodine" in our PICO format. At this point, we will not be using this as one of our search criteria as we are interested in all papers that discuss radioactive iodine exposure and cancer and do not want to potentially miss articles on this topic just because they do not mention that they have not compared patients who were exposed with patients who were not. Unfortunately, there is no way to be sure about whether or not some words should be used. It is always best to start with fewer terms so you get as broad a set of results as possible. After testing the search, it may well be that you will need to include more terms; but to begin with, three or four separate terms are all you should need. We now have three terms we can put into the database or databases of our choice. The next stage is to ensure that we have covered the various ways that these terms can be described.

Expanding and Improving Your Search Terminology

Every word or component that you draw out of your clinical question needs to be individually looked at and checked to see if there are multiple words (or synonyms) for it or if it needs to be pluralized. For example, if you are looking for articles about children you may also want to include the word 'child' in your search criteria. This idea can be expanded further so you could end up looking for children, child, infant, infants, baby, babies, teen, teens, teenager, teenagers, adolescent or adolescents.

There is no hard and fast rule about how many terms you should include. Common sense is the best guide in this process. There are obviously many different ways to look for 'children' but if you are looking for articles on vitamin C, you are reasonably limited in your choice of terms; ascorbic acid is the only other term you would use in your search. We also need to make sure we have the correct indexed or subject term for our search. We do this by looking up the thesaurus for the databases. In this case, we will use the Medical Subject Headings (MeSH) database. The MeSH database is the National Library of Medicine's (NLM) controlled vocabulary thesaurus used to describe journals articles contained in the PubMed database.

There are many important reasons to use a controlled vocabulary of terms in order to search the literature. The most important is that finding the correct term used by the indexers means that you will have a greater chance of finding as many relevant papers on your topic as possible. It will also increase the number of terms you will be searching for as controlled vocabularies allow for explosion of search terms. This means you will not only search for the term listed but all of the narrower terms that may fall under the main term. The MeSH database contains numerous potential synonyms that we can use to expand our search terminology. In this case, we start by looking for 'cancer'. This is a good term to use as an example as it covers the major reasons for using the MeSH database:

1. Locating the correct term
2. Identifying different synonyms for the term
3. Using the explode functionality to search for multiple terms

First, access the MeSH database at http://www.ncbi.nlm.nih.gov/mesh. Then, in the search box type in the word you are looking for. In this case, 'cancer'. The first thing you will notice is a list of 319 terms (at the time of writing). The term we are interested in is Neoplasms. Following this link through to its entry page, there are a number of points of information of interest to us. The first thing to look at is the description of the term. This is called a scope note and is found just beneath the heading. This description is what the NLM thinks the term Neoplasms means. It is important to read this to ensure that the term you have selected is accurate. In this case, the scope note reads: "New abnormal growth of tissue. Malignant neoplasms show a greater degree of anaplasia and have the properties of invasion and metastasis, compared to benign neoplasms" If you scroll down the page you will see the heading: Entry Terms. Entry terms are synonyms, alternate forms, and other closely related terms in a given MeSH record that are generally used interchangeably with the preferred term. So we see that under the entry terms is the word 'cancer'. Therefore we can now be certain that Neoplasms is the preferred terminology used by the NLM and it is the appropriate subject or MeSH term to use when searching the MEDLINE database.

Now that we have the correct term, we need to find the other various synonyms we may need to ensure our search is as broad as possible. The list of entry terms is the best place to look for these. Scanning through the list, we see the following:

- Neoplasm
- Tumors
- Tumor
- Cancer
- Cancers
- Benign Neoplasms
- Neoplasms, Benign
- Benign Neoplasm
- Neoplasm, Benign

From this list we select those we think should be used to increase our search terms. In this case the following terms could be selected: Neoplasm, Tumors, Tumor, Cancer and Cancers. As this is a US database, we should look for the English spelling of terms. So for Tumor (American spelling), we should make sure we also search for Tumour (English spelling).

We need to ensure that we do a subject heading, or MeSH, search so that our term is exploded. What the explosion of our search will do is to search for all of the narrower terms for Neoplasms. You can see the narrower terms by looking beneath the entry terms at the MeSH categories.

Under the Neoplasms term is Cysts, Hamartoma, Neoplasms by Histologic Type, Neoplasms by Site, Neoplasms Experimental, and so on. Under all of these terms are even narrower terms. So under Cysts we see Arachnoid Cysts, Bone Cysts, Branchioma, Breast Cyst, and so on again. There may be even more terms under these. So by exploding our term we not only search for our original term but all of the various and more specific terms at the same time.

There are various ways to ensure that you do a subject term search and this can change from database to database. There are many online tutorials available as well as help pages for each of them. You can consult these or you can seek assistance from your local librarian on how to effectively utilize a subject term search in each of the databases available. In this case, we will focus on how to do a subject search in the PubMed database.

Make sure you enter the term properly into the search box and type the word mesh or mh in square brackets following the term. They should look like this.

`Neoplasms[mesh] or Neoplasms[mh]`

The format is not case sensitive so capitalization is not needed. With the appropriate term found, the various synonyms identified and our search configured to explode the search term, it is time to put all facets together, ensuring that the Boolean operator 'OR' is between each term:

```
Neoplasms[Mesh] OR Neoplasms OR Neoplasm OR Tumors OR
Tumor OR Tumours OR Cancer OR Cancers
```

The word 'OR' does not need to be capitalized but it is quite useful to do so as it makes it easier to identify each individual word. The author has constructed hundreds of search strategies during his time as a librarian and has found that capitalizing the Boolean operators used not only helps keep track of the terms but also helps any research partners review the search. Our first search string is now ready to be entered into the database. The process above is repeated with each of the individual search components. Along with same refinement strategies discussed a later in the chapter, this gives us the following search strings.

```
"Iodine Radioisotopes"[Mesh] OR "Iodine Radioisotopes"
OR "Radioactive iodine" OR Radioiodine OR "radio-iodine" OR
"Iodine-131" OR "Iodine 131" OR RAI OR 131I
```

and

```
"Hyperthyroidism"[Mesh] OR "Graves Disease" OR Basedow
OR "Exophthalmic Goiter" OR "Exophthalmic Goiters"
OR hyperthyroidism OR thyrotoxicosis
```

If searching for more than one word, such as Graves disease, place quotation marks ("") around the words to ensure that the database searches for it as a phrase. A phrase search means that it will search for when the word Graves comes directly before the word disease. If you do not put quotation marks around the terms then the database may search for them as individual terms, this means papers that contain the word Graves and the word disease will be found irrespective of their location in the article, title or abstract.

With our three individual search strings completed, it is now time to run the search. We use the Boolean search operator 'AND' to combine our different search strings. Although you can combine them into one massive search string with 'AND' between them and brackets enclosing each individual concept, it is far wiser to use the advanced search features available in databases to do this. In the PubMed database, this is known as the PubMed Advanced Search Builder, although in other databases it is also commonly referred to as your search history. Once again look for online tutorials or seek the help of a librarian or other experienced user to help you do this. The concept is the same for each database. You enter and search for each component separately and then use the search builder or search history to do this.

In our case, we would search in the following way. Enter the search string

```
Neoplasms[Mesh] OR Neoplasms OR Neoplasm OR Tumors
OR Tumor OR Tumours OR Cancer OR Cancers
```

Click the search button. Then enter the second search string

```
"Iodine Radioisotopes"[Mesh] OR "Iodine Radioisotopes"
OR "Radioactive iodine" OR Radioiodine OR "radio-iodine"
OR "Iodine-131" OR "Iodine 131" OR RAI OR 131I
```

Click search again. Then enter the final search string

```
"Hyperthyroidism"[Mesh] OR "Graves Disease" OR Basedow
OR "Exophthalmic Goiter" OR "Exophthalmic Goiters"
OR hyperthyroidism OR thyrotoxicosis
```

Click search again. Go into the search builder and combine the searches with the 'AND' operator. Your search should then look like this.

Search	Query	Items found
#4	Search ((#1) AND #2) AND #3	1,109
#3	Search "Hyperthyroidism"[Mesh] OR "Graves Disease" OR Basedow OR "Exophthalmic Goiter" OR "Exophthalmic Goiters" OR hyperthyroidism OR thyrotoxicosis	43,627
#2	Search "Iodine Radioisotopes"[Mesh] OR "Iodine Radioisotopes" OR "Radioactive iodine" OR Radioiodine OR "radio-iodine" OR "Iodine-131" OR "Iodine 131" OR RAI OR 131I	60,974
#1	Search Neoplasms[Mesh] OR Neoplasms OR Neoplasm OR Tumors OR Tumor OR Tumours OR Cancer OR Cancers	2,940,477

Our initial search has now been completed and we have found 1,109 journal articles that contain something about cancer, radioactive iodine and benign thyroid disease. Now it is time for the next step in the search process.

Refine Your Search

Although the protocol of your review should be set up before doing any searching, this is only looking at correlations between populations, interventions and outcomes. The search you use does not have to be, and should not be, set in stone at the very outset. You need to build and run a search, as we have done above, check the results and look at the results you have found to help further improve your search.

Now is a good time to draw upon any personal expertise or the expertise of your review group. The first thing you need to do is to have your colleagues identify key articles or authors they may be aware of and ensure that your search has found these articles or authors. If key articles in the field have not been found, then you need to identify why not and modify your search in such a way as to capture them. Second, have your colleagues look over your search results and identify a few articles that they consider to be worth including in the review. You then need to look at the titles, abstracts and subject terms (MeSH terms) of these articles. See if there are any words or terms used in these articles that you have not included in your search strategy. If they are missing, then you will need to add them to the appropriate search string. For instance, in our case we may find that 'carcinoma' appears quite often in the literature and therefore we would add it to our cancer search string. Finally, have your colleagues look over your search terms and see if they have further suggestions to add to your search string. In our example above, the term 'radioiodine' and the acronym 'RAI' came from checking other papers and finding they were used in abstracts. Ensuring Graves disease was included as a keyword search came from the opinion of an expert.

Save Your Search Strategy

When searching, it is important to make sure you keep a record of the search you undertake. It is also useful to record changes you make as they occur and the reasoning behind the changes. As your search changes and is modified, it will be useful to report to any co-authors, why you made the changes and revisions that you did. Once your search is finalized and has been run, it is vitally important to make sure you have saved it. This can be done in most databases using the save search feature. But it is also very useful to copy and paste it into a word document.

Generally, the search occurs very early in the systematic review process and there is a lot of work done afterwards in reviewing the results, excluding the irrelevant articles, then extracting the data and running the meta-analysis. It could be 6 months or more before it is time to submit your article for publication and, if you do not have a copy of your search, you will be unable to report it in any real detail in your article. This will then damage the credibility of your article as other people will not be able to duplicate your methodology.

Select Your Databases

Now you have created a strong broad search and have saved a copy of it, you need to broaden your search to include other sources. This means searching for studies in additional databases. An important question is how many and which ones. The answer is as many as are relevant and you have the resources to search. Although there may be a desire to use as many databases as possible to maximize your coverage, the decision on how many and what databases you search should really be based on the resources you have available. For instance, if you are doing the project alone then there is not a lot of point searching ten databases as you will not have the capacity to deal with the number of results. By the same token, if you have a team of five people, then only searching MEDLINE will not be adequate.

The topic of your search is also a determining factor in what to use. If you are searching for surgical studies then you may want to search MEDLINE, Embase and CENTRAL. If you are looking for wound healing or patient care studies, then you should search the CINAHL database as well.

According to the *Cochrane Handbook for Systematic Reviews of Interventions*, you should search at least MEDLINE, Embase and CENTRAL when undertaking a systematic review. But dependent on the sort of information you are looking for, you may wish to search additional databases, dissertations or government web sites. The main databases that should be searched when undertaking a systematic review are:

- PubMed/MEDLINE: provides access to abstracts of the biomedical literature including research, clinical practice, administration, policy issues, and health care services. This database should always be searched if undertaking a systematic review.
- EMBASE: provides access to abstracts of biomedical and pharmacological journals. It has more coverage of European medical journals than MEDLINE. A good secondary database to search. Should be used in most instances when undertaking a systematic review.
- CENTRAL: the Cochrane Central Register of Controlled Clinical trials is a database containing all controlled trials found by the Cochrane Collaboration in MEDLINE, EMBASE, Review Groups, Specialized Registers and the Cochrane Hand Search Results Register. Should be searched especially if undertaking a systematic review of interventions.
- CINAHL: the Cumulative Index to Nursing and Allied Health Literature provides access to abstracts from nursing or allied health literature. Also covers materials from biomedicine, management, behavioural sciences, health sciences, librarianship, education and consumer health. This database needs to be included if undertaking a systematic review in nursing or allied health.
- PsycINFO: provides access to literature in psychology, psychiatry and related disciplines. Provides access to journals, books, reports, theses and dissertations. Needs to be included in your search strategy if looking at any topic regarding mental health.

Additional databases that can be searched are:

- Clinicaltrials.gov: developed to provide current information about clinical research studies, it can be used to identify unpublished studies and whether a study that is in progress could influence the findings of your systematic review.
- AMED: provides access to citations on alternative treatments to conventional medicine. Should be searched if conducting a systematic review about complementary or alternative medicine.
- Scopus: covers literature and quality Web sources in chemistry, physics, mathematics, engineering, life and health sciences, psychology, economics, social sciences, and biological and agricultural sciences.
- Web of Science: a multidisciplinary database that covers the literature from the sciences, social sciences, arts, and humanities.
- BIOSIS Previews: contains a lot of meetings and some medical journals. Controlled vocabulary is not suitable for medical searching.
- LILACS: health science literature published by Latin American and Caribbean authors.
- AIM: the African Index Medicus provides access to African health literature.
- Sociological Abstracts: provides coverage of sociological literature. Could be used when conducting reviews around community-based interventions.
- Current Contents Connect: covers research journals in the sciences, social sciences, and arts and humanities.

If searching for dissertations, there are three main places you can look.

- ProQuest Digital Dissertations: the primary search interface for dissertations. It is a subscription database, so you will need to check if your library has access to it.
- Google Scholar: not a very refined search interface and it does bring back many irrelevant results, but it has very broad coverage and is extremely useful, especially for those searchers who do not have access to a dedicated academic or hospital library that can afford a ProQuest Digital Dissertations subscription.
- WorldCat: provided by the Online Computer Library Center (OCLC) is a database of the holdings of a large number of libraries from around the world. Its coverage has a heavy western or English-speaking focus.

Searching for government reports can be a very time-consuming process, but a large amount of relevant information can be gained from them. The most effective way to search for government reports is to compile a list of relevant government departmental Web sites from the areas of the world you are interested in. Then use the Google advanced search feature to search a specific site or domain; for instance, if searching for reports by the World Health Organization (WHO), you would put www.who.int/ into the site or domain box in Google. To speed up the process, you should copy your search into a word document. You can then copy and paste the search into Google and modify the site you are searching. Searching this way, means you can cover quite a large number of Web sites in a reasonably short time frame.

Modifying Your Search Strategy

It is important to ensure you modify your search strategy so that it works in the various databases you are going to use. This means changing the subject terms in the controlled vocabulary; for instance, changing the MeSH term you have used in the MEDLINE database to an Emtree term, the controlled terms used in the Embase database. This is done in exactly the same way as finding a term. Sometimes the terms will be identical and sometimes they will be different. In this case, we will search for the MeSH term 'Neoplasms' in the Emtree controlled vocabulary.

Typing Neoplasms into the Emtree search box, we find that it returns the term Neoplasm. Not a huge difference but an important one in terms of our search structure. Now it is simply a matter of ensuring that the database searches for the Emtree term Neoplasm as well as the rest of our keywords. There are ways to do this in the database and once again it is suggested that you check for online tutorials, help pages or consult a search expert to learn how to do this effectively.

For the Embase.com search platform, the correct format for conducting a subject term, or Emtree, search, is as follows.

```
'neoplasm'/exp
```

This means it will search for the term neoplasm as an Emtree term and also explode the search, meaning we will capture all of the narrower terms that fall under it in the Emtree heading. So if you compare the two searches they will look like this.

- PubMed MEDLINE search:

```
Neoplasms[Mesh] OR Neoplasms OR Neoplasm OR Tumors OR
Tumor OR Tumours OR Cancer OR Cancers
```

- EMBASE search

```
'neoplasm'/exp OR Neoplasms OR Neoplasm OR Tumors OR Tumor
OR Tumours OR Cancer OR Cancers
```

If we do a similar thing in the CINAHL database we get the following formatted subject term search.

```
(MH "Neoplasms+")
```

This also means we will be searching for Neoplasms as a subject, or CINAHL headings, search and it will be exploded. If the database you wish to search does not include subject headings or a controlled vocabulary, such as Scopus, then remove the subject term from your search. Therefore a Scopus search would look like this.

```
Neoplasms OR Neoplasm OR Tumors OR Tumor OR Tumours OR
Cancer OR Cancers
```

Resolving Duplicate Results

Once your search is finalized and you have run it, you will find some duplicate results. This is due to the crossover in journal coverage in the databases. Manually checking for duplicates and removing them is a time-intensive process and a far easier way to check for and remove duplicates is to use bibliographic software such as EndNote. You can use the automated duplicate finding ability of bibliographic software to identify and remove the duplicate results in your search results. It is also a very useful tool to store the citations and full text of the articles that you have found. When the time comes to write your article, bibliographic software is also useful as it will look after your citations for you and automatically generate your reference list.

Hand Searching

Hand searching is the process of looking through the reference list of journal articles and reports and identifying any further articles of interest that you did not find during your electronic searching of the databases. The process is simple. Identify all of the relevant articles that you will be using to conduct your systematic review and meta-analysis. Then look through the reference list of each of them and identify any relevant articles that you have not yet identified. If there are any, you will need to access the full text versions and read them to see if they need to be included.

Reporting Your Search Strategies

Every systematic review should have the search strategies used incorporated into the article. This can be done by giving a brief overview of the key terms and databases used in your methodology and supplying the complete search structure in the appendix or as supplementary data. The search structure should, if possible, be copied and pasted directly from the database that you used. This means if you searched five databases you should have five searches in the appendix or supplementary material. If this is not possible due to space or word requirements, then you should copy and paste your primary search, in our case the PubMed search, and then provide information about how each search differed in each database. If we use our example, we would have:

PubMed search

Search	Query	Items found
#4	Search ((#1) AND #2) AND #3	1,109
#3	Search "Hyperthyroidism"[Mesh] OR "Graves Disease" OR Basedow OR "Exophthalmic Goiter" OR "Exophthalmic Goiters" OR hyperthyroidism OR thyrotoxicosis	43,627
#2	Search "Iodine Radioisotopes"[Mesh] OR "Iodine Radioisotopes" OR "Radioactive iodine" OR Radioiodine OR "radio-iodine" OR "Iodine-131" OR "Iodine 131" OR RAI OR 131I	60,974
#1	Search Neoplasms[Mesh] OR Neoplasms OR Neoplasm OR Tumors OR Tumor OR Tumours OR Cancer OR Cancers	2,940,477

The following search modifications were used in the different databases.

- EMBASE:

Changed the MeSH terms
`Neoplasms[Mesh], "Iodine Radioisotopes"[Mesh] and "Hyperthyroidism"[Mesh]`
to the Emtree terms
`'neoplasm'/exp, 'hyperthyroidism'/exp and 'radioactive iodine'/exp`
All other terms were included and remained the same.

- CINAHL:

Changed the MeSH terms
`Neoplasms[Mesh], "Iodine Radioisotopes"[Mesh] and "Hyperthyroidism"[Mesh]`
to the CINAHL headings
`(MH "Neoplasms+"), (MH "Iodine Radioisotopes") and (MH "Hyperthyroidism+")`
All other terms were included and remained the same.

- Web of Science and Scopus:

Removed the MeSH subject terms. All other terms were used and remained the same.

Anyone reading the above can see our primary search and replicate it exactly in the PubMed database. They can also then modify it reasonably easily with the same modifications that we used and run them in the same databases. This means they can exactly replicate our methodology and get identical results, which is one of the primary goals of a systematic review.

Selection of Studies and Data Abstraction

With the search done the next step is to look though all of the results you have found and decide which papers you need to keep and which can be discarded due to not meeting your criteria for inclusion. Although the process of determining inclusion and exclusion criteria and selecting which studies you should include is a complicated process. The following two basic questions cover all of the complexities of this process in a nutshell:

- Is the study relevant to the review's purpose?
- Is the study acceptable for review?

Using these questions as a guide will help you to work out the best criteria as to whether a study should or should not be included in your review. Inclusion and exclusion criteria for each individual review will be unique but the criteria will fall into the one of the following categories.

- Subjects or population involved
- Interventions being used
- Outcomes being of relevance
- Time period of the study
- Language of the study
- Methodological quality of the study

For the purposes of our systematic review and meta-analysis on radiodine we developed the following criteria. Studies were included if they reported on humans who had received radioiodine for diagnostic and therapeutic purposes, were prospective or retrospective cohort studies and compared cancer incidence in subjects that did and did not receive radioiodine. Studies were excluded if they used ^{131}I for treatment of thyroid cancers or for diagnostic purposes in thyroid cancer, studies that examined the relationship between cancer and ^{131}I exposure in nuclear accidents, descriptive studies, non-human studies, opinion papers, editorials and clinically irrelevant abstracts. With your criteria developed it is time to go through your results and select which you keep and which you discard. It is recommended that this process is undertaken by two independent people and then their results compared. A third person, preferably on the review team, should be called in if there are any discrepancies.

With a broad search across multiple databases you can expect to have a large number of results to go through, we found 2,929 studies in our search. When you consider that you can be expected to discard over 90 % of the studies you find it is unfeasible to obtain the full text of each individual paper. Therefore you should begin by excluding the obviously irrelevant studies. The following common steps should be followed when excluding studies.

Step 1: Discard duplicate results

If you searched more than a single database you will have found that the same studies have been found more than once. The first thing you should do is to remove these duplicates from your results. The best way to do this is to use a bibliographic software to do this automatically.

Step 2: Discard studies based on titles and abstracts

A great many of the studies you find will be obviously irrelevant and you should begin by going through and instantly discarding those studies by reading the title and if necessary the abstract. Look for common and easily determined reasons for exclusion, such as if they are written in a language that you will not be able to read or interpret, are an animal study (if you are after human studies) or are of a study type irrelevant to you (such as if it is a case series when you only want randomized controlled trials).

Step 3: Re-examine the titles and abstracts

After your first pass through your results revisit the remainder and spend more time analyzing the titles and abstracts. Now you remove the articles that with a slightly longer inspection of the titles and abstracts do not meet your criteria. If there is any doubt leave them in and make a decision later in the process.

Step 4: Retrieve the full text of the remaining studies

Now you will need to find the full text of any articles that have made it past the first three steps of the process. This can sometimes take time so do not think you need to wait until all the studies are available before moving to the next step.

Step 5: Exclude due to a failure to meet the criteria

With the full text of the article you can determine if the paper truly meets the exclusion and inclusion criteria you have set for your review. This can be a time-consuming process as excluding papers that made it this far into the process should be done carefully and you should keep a note of why each paper was excluded.

Step 6: Exclude due to other reasons

At this point you exclude all of the studies that although they meet the inclusion criteria and do not meet any of the exclusion criteria may need to be excluded for other reasons. This could be due to insufficient statistics for working out effect sizes or possibly studies which report incomplete or ambiguous results (reviewers may have sought further information from the authors before deciding to exclude it). It is important to record and report in your review why any paper was eventually excluded.

Throughout the entire process it is important to keep track of the numbers that were excluded at each step. Some reviews record the number of duplicate studies included and some do not. In our case we opted not to report the number of duplicates that we found. Commonly, steps 2 and 3 are combined into a single number. When reporting it is common practice to show the step by step process as a flow chart. See Fig. 12.1. The PRISMA flow diagram, available from their website, is growing in common usage and is a good format to follow when undertaking a systematic review. Next, you will need to extract the required data from the studies so that you can display summary findings in tables for your readers as well as run a meta-analysis.

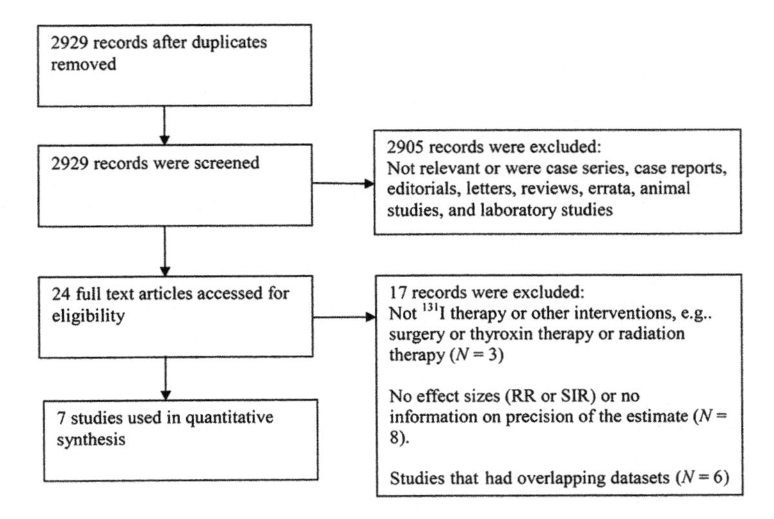

Fig. 12.1 Flow diagram of literature search for the systematic review/meta-analysis

What Data to Extract

There is a large potential body of data that could be extracted from an article, so it is important to have a firm understanding of what you want to extract once you have collected your studies. According to the Cochrane Collaboration, the data that will be collected should be outlined in your protocol to match the outcomes of interest in your systematic review. Although the specific data collected changes for each systematic review, the types of data fall into four broad categories. These are the study design and details, specific details about subjects, the treatments they were exposed to and the outcomes that were obtained. These categories can be remembered by the following acronym, **DOES**.

D: Design of study and other characteristics
O: Outcome data/results
E: Exposure/intervention and setting
S: Subject/participant characteristics

So long as your data extraction form, (covered later in this chapter), is designed to extract all relevant information in these four categories you should have collected all the necessary data you will need for your systematic review and subsequent meta-analysis. A list of data that can be collected based on the Committee on Standards for Systematic Reviews of Comparative Effectiveness Research is shown in Box 12.1.

The list of data that is required should be created with experts, such as clinical staff and statisticians, to ensure that all data of interest to the readers of the systematic review as well as all the data required to run the meta-analysis are collected.

Box 12.1.Types of Data Extracted from Individual Studies

General Information

1. Researcher performing data extraction
2. Date of data extraction
3. Identification features of the study:
    - Record number (to uniquely identify study)
    - Author
    - Article title
    - Citation
    - Type of publication (e.g. journal article, conference abstract)
    - Country of origin
    - Source of funding

D: Design of study and other characteristics

1. Aim/objectives of the study
2. Study design
3. Study inclusion and exclusion criteria
4. Recruitment procedures used (e.g. details of randomization, blinding)
5. Unit of allocation (e.g. participant, general practice, etc.)

S: Subject/participant Characteristics

1. Characteristics of participants at the beginning of the study, such as:
    - Age
    - Gender
    - Race/ethnicity
    - Socioeconomic status
    - Disease characteristics
    - Comorbidities
2. Number of participants in each characteristic category for intervention and comparison group(s) or mean/median characteristic values (record whether it is the number eligible, enrolled, or randomized that is reported in the study)

E: Exposure / Intervention and Setting

1. Setting in which the intervention is delivered
2. Description of the intervention(s) and control(s) (e.g. dose, route of administration, number of cycles, duration of cycle, care provider, how the intervention was developed, theoretical basis where relevant)
3. Description of co-interventions

O: Outcome Data/Results

1. Unit of assessment/analysis
2. Statistical techniques used
3. For each pre-specified outcome:
   - Whether reported
   - Definition used in study
   - Measurement tool or method used
   - Unit of measurement (if appropriate)
   - Length of follow-up, number and/or times of follow-up measurements
4. For all intervention group(s) and control group(s):
   - Number of participants enrolled
   - Number of participants included in the analysis
   - Number of withdrawals and exclusions lost to follow-up
   - Summary outcome data, e.g. dichotomous (number of events, number of participants), continuous (mean and standard deviation)
5. Type of analysis used in study (e.g. intention to treat, per protocol)
6. Results of study analysis, e.g. dichotomous (odds ratio, risk ratio and confidence intervals, *p* value), continuous (mean difference, confidence intervals)
7. If subgroup analysis is planned, the above information on outcome data or results will need to be extracted for each patient subgroup
8. Additional outcomes
9. Record details of any additional relevant outcomes reported
10. Costs
11. Resource use
12. Adverse events

Quality Assessment

During the data extraction process it is valuable to assess the methodological quality of each of the studies you will use (see Chap. 13). If a quality score is assigned, this should be reported in the data abstraction table. Usually the quality is assessed via a suitable methodological quality checklist. The checklist you use should be reported in your review and it is considered desirable to have two reviewers independently assign a quality score to each study with any discrepancies or disagreements resolved through a third reviewer if necessary.

There are many checklists available and if you use a pre-existing checklist you should ensure that it is customized to the specific needs of your review. The systematic review that was undertaken by the author used a modified version of a general checklist created by Doi and Thalib (Table 12.1). It was customized to

Table 12.1 Example of a methodological quality checklist for the meta-analysis of cancer risk after medical exposure to radioactive iodine in benign thyroid diseases

Items	Score
1. Was the data prospectively collected?	
0 = Retrospective cohort design	
1 = Prospective cohort study	
2. Were excluded (attrition) or inaccessible proportions < 20 %?	
1 = No	
0 = Yes or not reported	
3. Were the non-exposed 131-I cohort drawn from the same population base as the exposed cohort?	
0 = No or no description	
1 = Yes	
4. Were the distribution of risk factors for cancer between groups balanced/matched for (or adjusted for in the analysis):	
(1) Sex	
(2) Age at exposure	
(3) Latency (follow-up)	
(4) Any other factor balanced	
1 point for each balanced/adjusted item	
5. Was there a clear ascertainment of exposure of I-131 and were they clearly defined and precisely reported?	
1 = Yes, from hospital records	
0.5 = self reports from patients or no description	
6. Was there a clear ascertainment of cancer as an outcome and were it clearly defined and precisely reported?	
1 = Yes: diagnosis from any of following sources: physician, pathologists or hospital records	
0.5 = self reports from patients or not reported	
7. Was timing of cancer assessment adequate for cancer to occur?	
0: <5 years	
0.5: 5–10 years	
1: >10 years	
	Total = ______/10

include seven parameters reflecting the quality of the studies that were sought for this review: Prospective data collection, attrition (proportions excluded or inaccessible), selection bias, the distribution of risk factors for cancer between groups, ascertainment bias for both exposure and outcome, and duration of follow-up.

Data Collection Forms

To collate all the relevant data necessary from the original studies you have found into a format that you can use for your systematic review, a data extraction form is used. This is a form filled out by the data extractors that is standardized with all the necessary categories of data listed. Because all the required data has been decided beforehand, all the data extractors need is to find the necessary data in each of the studies and insert it into the appropriate place in the form. This means your data extractors do not need to be specialized in the field or qualified statistics experts.

The design of the form is straightforward and is generally a simple spreadsheet or paper form with the details of the article on the left hand side making up the rows and the data collected for each article on the top, making up the columns. To ensure that the data form works for the purpose of your systematic review, you should test it early in the review process with a small sample of relevant studies. This could save large amounts of time later. If the forms are found to be either too unwieldy, and therefore take too long to fill in, or do not cover enough data, this can pose serious problems if you are halfway through the data extraction. When designing the forms, you need to ensure you complete them in such a way that not only the experts involved in the review understand them but also that they can be easily understood by others who may be involved. It is unlikely that that the experts involved in the review will be the ones extracting the data so you need to ensure that other people can use the form. There is also the possibility that people will leave your review team or others may join at some stage in the process, so people not involved at the beginning of the review may be called upon to help extract data. Wherever possible, you should try to use standardized answers. This will not be possible in every instance but, for example, with yes/no questions (e.g. were the researchers blinded), do not leave a blank for people to fill out. Instead have a yes/no option where they can tick or select the answer. The decision on whether to use paper or electronic forms can come down to the personal preference of the data extractors. But it is recommended that paper only data extraction should be avoided due to the possibility of mistakes occurring during the transfer of the paper records into electronic format. Also paper forms add an extra step to the data extraction process, which can increase the time it takes. It is suggested that paper forms be used during the trial process of creating and testing the forms, and convert the form into an electronic format for the data extraction proper.

In our example "Cancer risk after medical exposure to radioactive iodine in benign thyroid diseases", it was determined that we required the following information to conduct a meta-analysis.

- Types of cancer reported
- Time frame of study
- Number of subjects included
- Outcome
- Years of follow-up
- Methods of ascertainment of outcome

Table 12.2 Data extraction table

Article details	Types of cancer	Time frame of study	Number of subjects	Outcome	Years of follow-up	Methods of ascertainmentof outcome	Reason for exposure to ^{131}I: diagnostic/ therapeutic	Activitiesof ^{131}I used (mean)	Comparison group	Exposed group
Dickman et al. (2003)	Thyroid	1952–1969	24,010	Standardized incidence ratios standardized for sex, attained age, and calendar year	27	Exposure: medical records Cancer: Swedish Cancer Register	Diagnostic	1.6 MBq	Patients examined with ^{131}I in seven Swedish hospitals from 1952 to 1969	Cancer incidence data matched for age, sex, region of residence, calendar year in Sweden from 1958 to 1969

- Reason for exposure to ^{131}I
- Number of exposed subjects in final cohort
- Activities of ^{131}I used (mean)
- Comparison group
- Exposed group

Therefore, a data extraction table was created (Table 12.2). Some of the data entered is free text and in these cases you have no choice but to allow the extractor to fill in the details as they see fit. But for others, you could have a selection of possible fields. For instance, in the reasons for exposure to ^{131}I column, there are only two possible responses: diagnostic or therapeutic. Therefore, by having only these two options available to the extractor reduces the possibility of inaccurate data being entered. An extreme example of the benefit of this approach is that the extractor may write ‘to determine if they had cancer’. This could be how it is referred to in the article but all that is needed for the purposes of the systematic review is for them to choose diagnostic. Another example of how to contain unwanted information is the choice of the words in the heading, “years of follow-up”. Although it could be assumed that, in this case, follow-up is measured in years, for people new to the research environment this is not necessarily intrinsic knowledge and you may end up with months being recorded instead. If this happens you can convert months to years but this is more time spent and more possibility for mistakes to occur. A well-designed form with the requirements of data extraction clearly outlined will save time later and decrease the potential for mistakes.

Missing Data

Missing data could cause problems in the accuracy of your review. Not all of the articles you find will report all the data that you want or need to undertake your analysis. The only real recourse in this situation is to e-mail the authors of the paper and ask them to supply the missing data. Obviously this request should be as polite as possible and it never hurts to acknowledge them in your article if they do supply the missing data, especially if it was hard to track down. If you cannot find the missing data, you will have to decide if you can include the study in your analysis or whether you need to exclude the article from your systematic review. For example, in the above scenario you could still include articles if they did not report on the activity of radioiodine administered but if they did not include the effect size for a subgroup of interest, you would need to exclude it.

Data Analysis, Results and Interpretation

Data analysis is the next step if data allows a meta-analysis to be performed (see Chaps. 14 and 15). Otherwise, extracted data is tabulated and a descriptive summary created for the results section. Findings are summarized based on the strength of the individual studies and the data abstracted so that conclusions may be reached based on best available scientific evidence from individual studies. All the aspects previously abstracted and summarized under the results section should be followed by a succinct discussion section focusing on strengths, weaknesses and applicability of the information to a health care setting. Recommendations regarding the intervention/exposure and implications for health care would also be important to discuss. Finally, it would be prudent to point out deficient areas to focus future research efforts and resources on.

Bibliography

Armstrong EC (1999) The well-built clinical question: the key to finding the best evidence efficiently. WMJ 98:25–28

Buscemi N, Hartling L, Vandermeer B, Tjosvold L, Klassen TP (2006) Single data extraction generated more errors than double data extraction in systematic reviews. J Clin Epidemiol 59:697–703

Chambers E (2004) An introduction to meta-analysis with articles from the Journal of Educational Research (1992–2002). J Educ Res 98:35–45

Corcoran J, Pillai VK, Littell JH (2008) Systematic reviews and meta-analysis. Oxford University Press, New York

Counsell C (1997) Formulating questions and locating primary studies for inclusion in systematic reviews. Ann Intern Med 127:380–387

Dickman PW, Holm LE, Lundell G, Boice JD, Hall P (2003) Thyroid cancer risk after thyroid examination with I-131: a population-based cohort study in Sweden. Int J Cancer 106:580–587

Doi SA, Thalib L (2008) A quality-effects model for meta-analysis. Epidemiology 19:94–100

Eden J (ed) (2011) Finding what works in health care: standards for systematic reviews. The National Academies Press for Committee on Standards for Systematic Reviews of Comparative Effectiveness Research, Institute of Medicine, National Academies Press, Washington, DC

Elamin MB, Flynn DN, Bassler D, Briel M, Alonso-Coello P, Karanicolas PJ, Guyatt GH, Malaga G, Furukawa TA, Kunz R, Schünemann H, Murad MH, Barbui C, Cipriani A, Montori VM (2009) Choice of data extraction tools for systematic reviews depends on resources and review complexity. J Clin Epidemiol 62:506–510

Higgins JPT, Green S (eds) (2011) Cochrane handbook for systematic reviews of interventions version 5.1.0 [updated March 2011]. The Cochrane collaboration. http://www.cochrane-handbook.org. Accessed 12 June 2012

Meline T (2006) Selecting studies for systematic review: inclusion and exclusion criteria. Contemp Issues Commun Sci Disord 33:21–27

Moher D, Liberati A, Tetzlaff J, Altman DG, The PRISMA group (2009) Preferred reporting items for systematic reviews and meta-analyses: the PRISMA statement. PLoS Med 6:e1000097. doi:10.1371/journal.pmed1000097

Reeves S, Koppel I, Barr H, Feeth D, Hammick M (2002) Twelve tips for undertaking a systematic review. Med Teach 24:358–363

US National Library of Medicine (2011) Medical subject headings, entry terms and other cross-references. http://www.nlm.nih.gov/mesh/intro_entry.html. Accessed 21 May 2011

US National Library of Medicine (2012) Medical subject headings, Neoplasms. http://www.ncbi.nlm.nih.gov/mesh/68009369. Accessed 21 May 2011

Chapter 13
Quality Assessment in Meta-analysis

Assessing the Validity of Study Outcomes

Maren Dreier

Abstract Quality assessment of primary studies to evaluate the reliability of study results is an essential and mandatory part of meta-analyses. It refers to the internal validity of a study and is described more precisely as assessing the risk of bias. Potential biases derive from selection of participants, data collection, analysis and selective reporting of study results. Quality assessment tools systematically collect information about study characteristics that may lead to bias in order to estimate the overall risk of bias. There are numerous tools available; they can be classified into checklists, scales and component ratings. Focusing on tools for assessing randomized controlled studies, an overview of covered elements of six selected generic tools is given. The Cochrane Collaboration's tool is described in more detail because it incorporates some important features. Practical aspects of conducting quality assessments are discussed including the meaning and importance of detailed and precise guidance.

Introduction to Quality Assessment

Quality assessment of primary studies to evaluate the reliability of the study results is an essential and mandatory part of meta-analyses. It has been shown that different levels of study quality are associated with different study results. Thus, systematic quality assessment is essential for clinicians and policy makers to make appropriate health care decisions without harming patients, wasting resources, and misleading future research.

No one standard exists for assessing study quality, as true results are unknown. Assessing study quality is a challenging process that needs rigorous methodological

---

M. Dreier (✉)
Hannover Medical School, Institute for Epidemiology, Social Medicine and Health Systems Research, Carl-Neuberg-Str. 1, 30625 Hannover, Germany
e-mail: dreier.maren@mh-hannover.de

S.A.R. Doi and G.M. Williams (eds.), *Methods of Clinical Epidemiology*, Springer Series on Epidemiology and Public Health,
DOI 10.1007/978-3-642-37131-8_13,

knowledge. A wide variety of tools are available for assessing study quality. This chapter describes the underlying rationale for quality assessments, the available tools and their application as well as the limitations of quality assessment.

The Definition of Study Quality

There are various definitions of the concept of study quality. Most regard study quality as internal validity defined as the extent of systematic distortion in the study results due to confounding or bias. For this reason, the term assessment of the risk of bias instead of study quality is more precise in this context, because a study may be performed to the highest possible quality standards, but nevertheless has an increased risk of bias. This may be the case, if, for example, blinding of participants is impossible as in studies that compare surgical versus non-surgical interventions. Here, we use the term study quality synonymously with risk of bias. The term methodological quality is also common in this context.

> Quality assessment in the context of meta-analysis is described more precisely as assessing the risk of bias and refers to the internal validity of a study.

Study Quality Versus Reporting Quality

Towards the end of the 1990s, the first checklist for reporting quality in randomized controlled trials (RCT), CONSORT, was published, followed by further design-specific lists such as STARD, STROBE and PRISMA (see http://www.equator-network.org) aiming to enhance the reporting quality, which was often criticized as being inadequate. The target group for these checklists include authors, reviewers, editors and readers of published studies.

Adequate comprehensive reporting of a study is a necessary prerequisite for assessing the study quality. The reporting quality of a trial describes whether information about the design, the conduct and the analysis is available and complete in the original paper without assessing the resulting implications for the validity of the study results. Reporting quality addresses, for example, if loss to follow-up is described in detail (number/proportion/characteristics of loss to follow-up). In terms of quality assessment on the other hand, assessment of the appropriateness of the loss to follow-up with regard to the risk of bias (the extent of loss to follow-up, differential loss to follow-up) will be looked at. Consequently, studies with equal reporting quality may result in huge differences in the risk of bias. Nevertheless, numerous quality assessment tools are mixed up with items of reporting quality, and that can lead to a misinterpretation of the study quality.

Quality assessment is based on adequate reporting quality. Relevant unreported methodological aspects cannot contribute to the quality assessment, and thus the study quality may be systematically underestimated. To solve this problem, contacting the authors of original papers to obtain missing information to gain greater accuracy in quality assessment seems to be a good idea at first sight, but should be done very cautiously, because the answers may be too favourable and difficult to verify.

Adequate reporting quality is a prerequisite for assessing the risk of bias. However, reporting quality should never serve as a surrogate for the risk of bias.

The Rationale of Quality Assessments

As the true answer to a study question is unknown, the aim is to estimate the reliability of the study results by using specific study characteristics that are supposed to affect internal validity. Several studies have shown that the results of meta-analyses are associated with the quality attributed to the studies included suggesting that more biased studies show (more) deviant results from the true results. The observed effects varied. Moher et al. (1999) demonstrated that high quality studies result in lesser effects of interventions compared with low quality studies, while Odgaard-Jensen et al. (2011) demonstrated that effect sizes can change in either direction but again generally in the direction of larger estimates of effect for lower quality studies. Thus, a meta-analysis with many low quality studies may lead to an overestimation of the effect and may, therefore, have an undesirable impact on clinical practice and health policy decisions.

To minimize the extent of bias in study results, several methodological safeguards, such as randomization and allocation concealment, have been developed. These are described in detail in the following sections. Quality assessment tools are designed to systematically collect information about study characteristics that may lead to bias in order to estimate the overall risk of bias.

The results of the quality assessments can be incorporated in different ways into meta-analyses. One strategy is to define a threshold for the inclusion of studies by restricting the analysis to studies with a low risk of bias. Another method is to stratify by study quality or perform a sensitivity analysis by first including only studies with a low risk of bias and subsequently adding the others. Finally an overall quality score can be computed that can be used for comparison across studies but has to be utilized carefully – see below.

Table 13.1 Principles of internal validity and its potential threats

Principles of internal validity	Aims	Methods	Potential threats
Equivalence of study groups (avoiding selection bias and confounding)	Study groups do not differ in their known and unknown baseline characteristics	Random sequence generation	Selection bias
		Allocation concealment	Attrition bias (Exclusion bias)
		Intention-to-treat analysis[a]	
Equivalence of performance (avoiding information bias)	Preferably no recognizable differences in intervention, and no differences in co-interventions, care and setting	Blinding of participants and personnel	Performance bias
		Standardization	Attrition bias
Equivalence of data collection process (avoiding information bias)	Identical assessment of baseline characteristics, outcomes and adverse effects in all study groups	Blinding of participants, personnel and outcome assessors	Detection/ measurement bias
		Standardization	Attrition bias
		Training of assessors	

[a]This is not true for non-inferiority or equivalence studies

Bias Affecting Internal Validity

Internal validity refers to the extent to which study results are true. Results can differ from the true effect because of random or systematic errors. Random errors affect the precision of the study results estimated with confidence intervals and *P* values, and depend on the sample size and the variation in measurements. Systematic errors such as bias or confounding can arise from how the study is conducted (selection of participants, performance, data collection and analysis, as well as selective publication) and lead to results that systematically over- or underestimate the true relationship between an intervention and its specific outcome. Quality assessment focuses on the potential threats from bias. Thus, it is important to distinguish between random and systematic errors.

The main principles to produce highly valid study results are equivalence of study groups, performance and data collection (Table 13.1). In the following, the most important sources of bias and the corresponding methods to avoid them are described.

Confounding and Selection Bias

Imbalanced Study Groups

Ideally, study groups should differ only in the intervention of interest and not in other personal or environmental characteristics, especially not in (potential) prognostic factors such as age, sex or stage of disease, to ensure that study results can be attributed causally to differences between intervention and control conditions (and not, for example, to the intervention group being younger than the control group). Differences in the characteristics of the study groups can give rise to confounding. The gold standard to avoid baseline imbalances is randomization with each participant having an equal chance of being assigned to one of the study groups and the allocation not being predictable. A successful randomization results in equivalence of the study groups not only of known but also of unknown characteristics and can be corroborated by equal baseline characteristics of the study groups.

The randomization process can be divided into two parts: random sequence generation and its unbiased implementation, which is assured by allocation concealment. Concealing the allocation sequence from those involved in enrolment avoids breaking of the randomization code. Otherwise, sicker patients could, for example, be chosen for the new treatment more frequently because the allocating doctor expects better effects for them, which may lead to underestimation of the true effect.

Attrition Bias

It is very important that the study groups remain as complete as possible throughout the study with little or no withdrawals or participants lost to follow-up. This is essential, because the study participants who withdraw may be different from the remaining participants; for example, they may be sicker or benefit less from the intervention, which may lead to biased study results. An attrition bias can also result from excluding participants despite available outcome data (exclusion bias).

To avoid biased study results, the analyses should primarily follow the intention-to-treat (ITT) principle, which analyses all participants according to their randomly assigned group, regardless of whether they were compliant with the study protocol or not. The main reason for ITT analyses is to maintain the randomized study groups, although the true effects are likely to be underestimated. Nevertheless, ITT analyses do not minimize bias from missing values in study outcomes. Methods to minimize loss to follow-up are important throughout the study. Even participants who give up their assigned intervention should be assessed for outcome variables until the end of the study to avoid missing values. An important method to avoid a differential loss to follow-up causing an attrition bias is blinding, for example, blinding the participants may avoid participants in the control group dropping out because they do not expect any benefit, and blinding of the personnel may ensure that they make equal efforts to convince patients to remain in the study.

Information Bias

Performance Bias

Performance bias results from systematic differences between the study groups in the care that is provided and the setting where the study takes place, for example, different monitoring, co-interventions, and diagnostic procedures. The main methods to avoid performance bias include blinding participants and study personnel to make sure that the allocated intervention remains unknown, and thus has no effect on the participants' and study personnel's behaviour.

Detection Bias

Detection bias, also called measurement bias, can arise from systematic differences in the outcome assessment between the study groups, especially, if subjective end points will be collected. This type of bias can be minimized by using standardized methods and blinding of both the participants and the outcome assessors. For some interventions, the intended blinding may not be effective due to (side) effects, such as lowering of the heart rate caused by beta-blockers or a specific taste of a drug. This may be avoided by comparing drugs with similar side effects or using an active placebo. However, guessing the allocated study group correctly can simply reflect a better outcome of the intervention of interest. Ineffective blinding can also induce performance bias.

Other Sources of Bias

Selective Reporting Bias

More recently, research has focused not only on bias arising from trial conduct but from publication of studies. There is increasing empirical evidence that within one published study, the outcomes may be reported depending on the nature and direction of the results. This selective reporting bias, also called outcome reporting bias, can be viewed as a kind of publication bias (defined as selective reporting of entire trials, see Chap. 15) within a study but not on study level. Others define both selective reporting bias and publication bias among others as reporting bias.

Examples of selective reporting include unreported harms, negative results presented more favourably or conclusions not supported by the actual study results. Moreover, it was found that in a substantial number of studies, primary outcomes were newly added, left out, or modified compared with the study protocol. However, in practice the study protocol may not be available for every study, making it difficult to address selective reporting in the quality assessment.

Conflict of Interest

Authors' conflict of interests is of growing major concern because there is evidence that it can lead to biased study results. For example, reviews that do not find an association between second-hand smoking and health risks are more often written by authors with ties to the tobacco industry. Studies in the field of dental research showed more favourable results in industry-sponsored studies. There is also evidence suggesting that pharmaceutical-sponsored drug studies are more likely to report outcomes in favour of the sponsored product than non-industry-funded studies. This could not be explained by the reported methodological quality, but by inappropriate control interventions and publication bias.

It is not clear how to best deal with potential conflicts of interest. Simply excluding industry-sponsored studies may not be practicable, especially not for drug studies because those are often sponsored by industry. Efforts are focused on higher transparency and decreasing the sponsor's influence on study conduct and publication.

Overview of Quality Assessment Tools

There are numerous tools to assess the study quality that show great variability in the development process, the number of items included, time and effort required and content characteristics. Tools can be divided into generic versus topic-specific tools for specific diseases or health technologies. Most of the tools have been developed for a specific design. This chapter focuses on tools for assessing randomized controlled trials (RCT), which are considered the gold standard for studies investigating the effectiveness or efficacy of different preventive or therapeutic interventions. In addition, diagnostic studies and screening studies are prone to additional specific bias affecting the validity of the study results and have specific checklists, for example, QUADAS for diagnostic studies suggested by Whiting et al. (2004).

Types of Tools

The tools can be classified into checklists, scales and component ratings (Table 13.2). Checklists and component ratings are qualitative tools, whereas scales provide a quantitative assessment. The first checklist was published in 1961 followed by nine others before1993. The first scale was published in 1981 and a total of 24 before 1993. Component ratings have been developed more recently; one well-known tool, the risk of bias tool, was introduced by the Cochrane Collaboration in 2008.

Quantitative Tools: Scales

In scales, each item is attributed a numerical score that will be combined to provide a sum score. Adding the single scores leads to an implicit weighting of the single items with each item being assigned the same weight. Some scales include

Table 13.2 Types of tools

Type of tool	Definition	Examples of tools
Scale	Every item is assigned a numerical score combined to a summary score	Six-Item scale of Jadad et al. (1996) Scale of Downs and Black (1998)
Checklist	A list of at least two items without a numerical rating system	Critical appraisal for therapy articles from the Centre for Evidence Based Medicine (CEBM) MERGE (Method for Evaluating Research and Guideline Evidence) checklist for RCT (Liddle1996)
Component rating	Components such as randomization or blinding are assessed without a numerical rating	Cochrane Collaboration's risk of bias tool (Higgins et al. 2011a)

weighting methods derived from the attributed importance of the single items for the internal validity. Two of the best known scales were developed by Jadad et al. (1996) and Downs and Black (1998). The short six-item scale of Jadad et al. covers mainly aspects of the reporting quality rather than of conduct and thus cannot be recommended for a thorough quality assessment. The scale of Downs and Black includes elements that rate the internal validity as well as the external validity and reporting, and addresses RCTs and observational studies (see Table 13.3).

Using a summary score to characterize the level of study quality may be tempting by its simplicity: different studies can be easily compared with only one parameter that can also be used as a threshold for the inclusion of studies, as a weighting factor or for stratifying the studies in the meta-analysis. However, the main argument against using scales is that an empirical proof for the implicit or explicit weighting of the components included is missing, but this might now have been addressed (see Chap. 14).

Qualitative Tools: Checklists and Components Ratings

A checklist is a list with at least two items without a numerical scoring system. Tools with a component rating, also called domain-based evaluation, include components or domains such as randomization or blinding that were rated qualitatively. The single components usually include several single methodological elements and thus need detailed and precise guidance to achieve an unambiguous assessment.

Some qualitative tools provide a qualitative rating of groups of several items resulting in a qualitative overall assessment, or do a qualitative overall assessment defined by specific criteria in the first place. There is great variety in doing and incorporating summary assessments. However, there is no evidence that qualitative overall assessment of study quality is associated with the study results.

Table 13.3 Covered elements of selected generic quality assessment tools with guidance

	[a]D&B	[b]Cochrane	[c]EBHPP	[d]CASP	[e]LBI	[f]MERGE	[g]CEBM
Type of tool	Sc	Co	Co	Cl	Cl	Cl	Cl
Number of items/components	27	6	6	10	9	10	7
Overall assessment	NSS	QCR QOC	NSS	–	QOR	QOR	–
Randomization	●	●	●	●	●	●	●
Allocation concealment	●	●			●	●	
Comparable groups		●	●	●	●	●	●
Blinding of participants	●	●	●	●	●	●	●
Blinding of outcome assessors	●	●	●	●	●	●	●
Blinding of personnel		●		●	●		
Equivalence of performance				●		●	●
Placebo comparable with verum				●			
Co-interventions			●				
Contamination	●		●				
Compliance	●		●				
Valid methods	●						
Reliable methods	●					●	
Follow-up							●
Follow-up contemporary	●						
Adequate analyses	●			●		●	●
ITT analysis	●	●	●		●	●	
Missing values		●	●				
Confounding	●			●			●
Loss to follow-up	●	●	●		●	●	●
Differential loss to follow-up					●		
Selective reporting	●	●	●	●			
Financial funding							

Cl checklist, *Co* component rating tool, *NSS* numerical sum score, *QCR* qualitative rating on level of components, *QOR* qualitative overall rating, *Sc* scale

[a]Downs and Black, UK, 1998. For RCTs and observational studies, before and after studies

[b]Risk of bias tool (2nd version) from the Cochrane Collaboration, 2011. Further elements than those marked may be covered by a component for other bias

[c]Effective Public Health Practice Project (EPHPP): quality assessment tool from MacMaster University, Canada, 2004

[d]Critical Appraisal Skills Programme (CASP) from the Public Health Resource Unit, UK, 2006

[e]Ludwig-Boltzmann Institute, Austria, 2007. German language

[f]MERGE (Method for Evaluating Research and Guideline Evidence) checklist for RCT

[g]Critical Appraisal for Therapy Articles from the Centre for Evidence Based Medicine (CEBM), University of Oxford, UK, 2005

The Cochrane Collaboration's Risk of Bias Tool

The Cochrane Collaboration's tool is described in more detail as an example of a component rating because it is widely discussed and incorporates some important features. In 2008, the Cochrane Collaboration published a new tool developed in expert meetings to harmonize quality assessments in Cochrane reviews. This risk of bias tool became the recommended method throughout the Cochrane Collaboration. The first version was followed by an evaluation by Higgins et al. (2011) resulting in slight differences. The tool includes six bias domains with a total of seven sources of potential bias: random sequence generation, allocation concealment, blinding of participants and personnel, blinding of outcome assessors, incomplete outcome data, selective reporting, and anything else, ideally prespecified. Each of these items has to be assigned a low, high or unclear risk of bias with the reasons for the judgement being documented to make the assessment process transparent. Items that are not sufficiently reported to be judged are rated as having an unclear risk of bias. Detailed guidance for making judgements is provided in the Cochrane Handbook and described by Higgins et al. (2011a).

As the assessment of the risk of bias is based partly on judgements that include much room for interpretation, the reported inter-rater reliabilities are not as high as for other tools. This can partly be explained by not giving exact thresholds for a high risk rating, (e.g. more than 15 % missing values for rating the component "incomplete outcome data"), because there is no evidence for these thresholds, and thus, the respective items need more subjective assessment. The relatively high amount of subjective judgement in the risk of bias tool remains challenging for the reviewers and requires profound methodological knowledge.

If multiple outcomes are examined in meta-analysis, one important new principle applied in the risk of bias tool involves an outcome-specific evaluation of the blinding items and missing outcome data. Whereas the risk of bias of the reported methods of sequence generation and allocation concealment have an impact on all study outcomes, the effect of blinding on the risk of bias depends on the subjectivity of the specific outcome and can be assumed to be lower in more objective outcomes. The outcome-specific assessment takes into account that different outcomes incorporate different levels of risk of bias, for example, the risk of bias of not blinding outcome assessors of all-cause mortality may be judged as low in contrast to the risk of bias of self-reported outcomes. Missing values may vary between different outcome variables, for example, at different points of time, and thus lead to different judgements of the resulting risk of bias.

The overall risk of bias within a study is estimated qualitatively and should be given for each important outcome based on the magnitude and the direction of bias. Depending on the research topic, the judgement is focused on the components defined as most important, also called key domains. A low risk of bias in all key domains results in an overall low risk of bias that is unlikely to alter the study results seriously, whereas a high risk of bias for one or more key domains leads to an overall high risk of bias that may affect the results considerably.

Elements of Quality Assessment Tools

Moher et al. (1995) and West et al. (2002) have compared available tools with respect to their formal and methodological characteristics and content. Based on these reviews, study elements were selected that have either demonstrated empirical evidence of an effect on the study results, such as randomization, allocation concealment, blinding, ITT analysis, selective reporting bias, or its distorting effect (see studies by Kjaergard, Odgaard-Jensen, Pildal, Wood and Porta in the bibliography for more information). Table 13.3 provides an overview of the elements covered in several generic tools that also provide guidance for the reviewer on how to rate the items or components (derived from Dreier et al. (2010)). Table 13.3 raises no claim on selection of the best tools or on completeness. The following elements are considered to characterize the tools: randomization method, allocation concealment, comparability of study groups at baseline, blinding of the participants, outcome assessors and study personnel, equivalence of performance except of the intervention, placebo comparable with verum, co-interventions, contamination, compliance, statistical methods, ITT analysis, incomplete outcome data, selective reporting, and source of funding.

The component rating tool from the Cochrane Collaboration has already been described. As it provides a component for other bias, further elements than those indicated may be covered. The Effective Public Health Practice Project (EPHPP) Quality Assessment Tool can be used to evaluate not only RCTs but also observational studies. It covers six components rated as strong, moderate or weak derived from a numerical scoring of the single items that is averaged across the components to the final sum score. Based on their sum score, studies are assessed as having a weak, moderate or strong quality. This numerical scoring characterizes this tool as a quantitative tool. The evidence that scales do not correctly rate the study quality, as well as the fact that there is no rationale for the implicit weighting of elements, also apply for the EPHPP tool. However, without the recommended numerical rating, the tool can be classified as a component rating system.

The scale by Downs and Black (1998) was developed for assessing RCTs and observational studies and includes explicitly designated items of internal validity, external validity and reporting quality. Like for all scales, using the tool without numerical rating and restricting to items of internal validity is recommended.

The other tools in the table are checklists; two of them provide a qualitative overall assessment. The MERGE checklist for RCTs has been developed by the New South Wales Department of Health, which also provides further design-specific checklists. It is used by the Scottish Intercollegiate Guidelines Network and is available in *A Guidelines Developer's Handbook*. Each item can be rated as well covered, adequately addressed, poorly addressed, not addressed, not reported, or not applicable. An overall assessment of how well the study was done to minimize bias has to be coded "++", "+", or "−", as well as the likely direction in which bias might affect the study results. Another checklist developed by the Ludwig-Boltzmann Institute, Austria, is available in German in a manual on doing health technology assessments. The items can be rated as yes, no, not reported, not applicable resulting in an overall assessment of internal validity being good, sufficient, or insufficient. Two further checklists, one from the Centre of Evidence Based Medicine, and another one from the Critical Appraisal Skills Programme, provide no explicit overall assessment. The Critical Appraisal for Therapy articles assesses the internal validity of the study results. Each item can be rated as yes, no, or unclear; free text is provided for additional comments. The tool of the Critical Appraisal Skills Programme starts with two screening questions to restrict the assessment to RCTs. From the remaining eight items, four address not the risk of bias, but random errors and external validity.

Selecting a Tool for Assessing Study Quality

The choice of a quality assessment tool should be based on how comprehensively a tool covers the important elements for the specific research topic. Tools with a numerical scoring system should not be used without a proper quantitative framework (see Chap. 14). The elements of the tool should not be mixed up with parameters to assess reporting quality or external validity. An accompanying guide with details on the operationalizations of the items or components is recommended to support the rating. It might be practical to select a generic tool that can be applied for different meta-analyses, and to adjust it to the specific requirements of the respective research topic and studies.

Conducting the Quality Assessment

Assessing study quality is a subjective process that is prone to reviewer bias and may lead to systematically distorted results. It is generally accepted that two reviewers should independently assess study quality and that discrepant ratings be solved by consensus, preferably including at least one additional person. In some reviews, the first reviewer rates and a second reviewer confirms these ratings. However, as Buscemi et al. (2006) demonstrates that in? data extraction, the latter method, although time-saving, produces more errors compared with both reviewers first rating the risk of bias independently of each other. There are no such studies for quality assessment, but since quality assessment is typically more subjective than data extraction, the advantage of independent concurrent ratings could be even greater.

Blinding reviewers to avoid reviewer bias means that the reviewers assess the studies without knowing any identifying characteristics (such as title, authors, and journal). Nevertheless, reviewer blinding is not common and may be impossible in cases where reviewers have profound knowledge of the review topic. The effectiveness of reviewer blinding is not clear because empirical studies examining the effect of reviewer blinding on quality assessment by Clark et al. (1999), Jadad et al. (1999), Moher et al. (1998) and Zaza et al. (2000) show inconsistent results.

Further methods to minimize the subjectivity of evaluations include standardization of the process to assure a high agreement in ratings. This can be measured by the inter-rater reliability. Oxman (1994) has shown that higher agreement is usually prevalent in items that provide less room for interpretation. For example, for the Cochrane Collaboration tool, it was shown by Hartling et al. (2009) that the inter-rater reliability was lower for components that need more subjective ratings. This finding supports the need for a detailed guidance on how to rate the single items and/or reviewer training in order to achieve consistent ratings. Preferably, the chosen tool should be piloted in advance to determine if the guidance for the assessment needs to be supplemented or modified to make sure that criteria are applied consistently. Three to six studies that cover a broader range of study quality are recommended for the piloting process.

A quality assessment tool should be used together with a detailed and precise guide to minimize the subjectivity of the rating and to ensure that reviewers know exactly how to rate the items or components. A piloting process is recommended to adjust the guide to the studies that are being analysed.

A recently introduced principle involves outcome-specific assessment of the risk of bias for the items blinding and incomplete outcome data. These items can have different impacts on the risk of bias depending on the specific outcome. An outcome-specific assessment, at least for groups of outcomes, is explicitly required in the Cochrane Collaboration's tool. This useful principle might also be implemented when using other tools.

When assessing multiple outcomes, the risk of bias should be rated separately for each outcome (outcome-specific risk of bias assessment).

Derived from the Cochrane Collaboration's tool, it is recommended that the study characteristics that were used for judgements are documented. Although this approach may be time consuming, it may hasten consensus, and importantly, ensure the transparency of the quality assessment.

In summary, reviewers should have rigorous methodological knowledge. It is useful to include experts on the review topic (with sufficient methodological knowledge) to perform the quality assessment or to advise on the clinical aspects and specific methodological problems. Some analytical methods may go beyond the reviewers' expertise; in those cases it may be necessary to recruit additional experts to assess the appropriateness of complex statistical analyses, for example, multivariable models and other methods used in special study designs.

Quality assessment is a lengthy process that includes the preparation (choice of tool, the operationalization process of the single items or components, and a priori decision on how to integrate the results of the quality assessment), the actual rating, and formulation of the final consensus. Therefore, for reliable quality assessment being a key part in meta-analysis with far-reaching implications on health care decisions, it is crucial to have realistic time expectations and sufficient resources and personnel.

Bibliography

Abraham NS, Moayyedi P, Daniels B, Veldhuyzen van Zanten SJ (2004) Systematic review: the methodological quality of trials affects estimates of treatment efficacy in functional (non-ulcer) dyspepsia. Aliment Pharmacol Ther 19:631–641

Al Khalaf MM, Thalib L, Doi SA (2011) Combining heterogenous studies using the random-effects model is a mistake and leads to inconclusive meta-analyses. J Clin Epidemiol 64:119–123

Armijo-Olivo S, Stiles CR, Hagen NA, Biondo PD, Cummings GG (2012) Assessment of study quality for systematic reviews: a comparison of the Cochrane collaboration risk of bias tool and the effective public health practice project quality assessment tool: methodological research. J Eval Clin Pract 18:12–18

Balevi B (2011) Industry sponsored research may report more favourable outcomes. Evid Based Dent 12:5–6

Barnes DE, Bero LA (1998) Why review articles on the health effects of passive smoking reach different conclusions. JAMA 279:1566–1570

Begg C, Cho M, Eastwood S, Horton R, Moher D, Olkin I, Pitkin R, Rennie D, Schulz KF, Simel D, Stroup DF (1996) Improving the quality of reporting of randomized controlled trials. The CONSORT statement. JAMA 276:637–639

Buscemi N, Hartling L, Vandermeer B, Tjosvold L, Klassen TP (2006) Single data extraction generated more errors than double data extraction in systematic reviews. J Clin Epidemiol 59:697–703

Campbell MK, Elbourne DR, Altman DG (2004) CONSORT statement. Extension to cluster randomised trials. BMJ 328:702–708

Chan AW, Krleza-Jeric K, Schmid I, Altman DG (2004) Outcome reporting bias in randomized trials funded by the Canadian Institutes of Health Research. Can Med Assoc J 171:735–740

Cho MK, Bero LA (1994) Instruments for assessing the quality of drug studies published in the medical literature. JAMA 272:101–104

Clark HD, Wells GA, Huet C, McAlister FA, Salmi LR, Fergusson D, Laupacis A (1999) Assessing the quality of randomized trials. Reliability of the Jadad scale. Control Clin Trials 20:448–452

Downs SH, Black N (1998) The feasibility of creating a checklist for the assessment of the methodological quality both of randomised and non-randomised studies of health care interventions. J Epidemiol Community Health 52:377–384

Dreier M, Borutta B, Stahmeyer J, Krauth C, Walter U (2010) Comparison of tools for assessing the methodological quality of primary and secondary studies in health technology assessment reports in Germany. GMS Health Technol Assess 6:Doc07

Dwan K, Altman DG, Arnaiz JA, Bloom J, Chan AW, Cronin E, Decullier E, Easterbrook PJ, Von Elm E, Gamble C, Ghersi D, Ioannidis JP, Simes J, Williamson PR (2008) Systematic review of the empirical evidence of study publication bias and outcome reporting bias. PLoS One 3: e3081

Greenland S (1994) Invited commentary. A critical look at some popular meta-analytic methods. Am J Epidemiol 140:290–296
Haahr MT, Hrobjartsson A (2006) Who is blinded in randomized clinical trials? A study of 200 trials and a survey of authors. Clin Trials 3:360–365
Hartling L, Ospina M, Liang Y, Dryden DM, Hooton N, Krebs SJ, Klassen TP (2009) Risk of bias versus quality assessment of randomised controlled trials. Cross sectional study. BMJ 339: b4012
Herbison P, Hay-Smith J, Gillespie WJ (2006) Adjustment of meta-analyses on the basis of quality scores should be abandoned. J Clin Epidemiol 59:1249–1256
Higgins JP, Altman DG, Gotzsche PC, Juni P, Moher D, Oxman AD, Savovic J, Schulz KF, Weeks L, Sterne JA (2011a) The Cochrane collaboration's tool for assessing risk of bias in randomised trials. BMJ 343:d5928
Higgins JPT, Altman DG, Sterne JAC (2011) Assessing risk of bias in included studies. In: Higgins JPT, Green S (eds). Cochrane handbook for systematic reviews of interventions. Version 5.1.0 (updated March 2011). The Cochrane Collaboration. Available from http://www.cochrane-handbook.org
Huwiler-Müntener K, Jüni P, Junker C, Egger M (2002) Quality of reporting of randomized trials as a measure of methodologic quality. JAMA 287:2801–2804
Jadad AR, Moore RA, Carroll D, Jenkinson C, Reynolds DJ, Gavaghan DJ, McQuay HJ (1996) Assessing the quality of reports of randomized clinical trials. Is blinding necessary? Control Clin Trials 17:1–12
Juni P, Witschi A, Bloch R, Egger M (1999) The hazards of scoring the quality of clinical trials for meta-analysis. JAMA 282:1054–1060
Juni P, Altman DG, Egger M (2001) Systematic reviews in health care: assessing the quality of controlled clinical trials. BMJ 323:42–46
Kjaergaard LL, Villumsen J, Cluud C (2001) Reported methodologic quality and discrepancies between large and small randomized trials in meta-analyses. Ann Intern Med 135:982–989
Kunz R, Oxman AD (1998) The unpredictability paradox. Review of empirical comparisons of randomised and non-randomised clinical trials. BMJ 317:1185–1190
Lexchin J, Bero LA, Djulbegovic B, Clark O (2003) Pharmaceutical industry sponsorship and research outcome and quality: systematic review. BMJ 326:1167–1170
Liddle J, Williamson M, Irwig L (1996) Method for evaluating research and guideline evidence. New South Wales Department of Health, Sydney
Lohr KN, Carey TS (1999) Assessing "best evidence". Issues in grading the quality of studies for systematic reviews. Jt Comm J Qual Improv 25:470–479
Ludwig-Boltzmann Institut HTA. (Internes) Manual–Abläufe und Methoden Tl. 2. 2007. Wien, Ludwig Bolzmann Institut, Health Technology Assessment. HTA-Projektbericht, Nr. 006
Mathieu S, Boutron I, Moher D, Altman DG, Ravaud P (2009) Comparison of registered and published primary outcomes in randomized controlled trials. JAMA 302:977–984
McGauran N, Wieseler B, Kreis J, Schuler YB, Kolsch H, Kaiser T (2010) Reporting bias in medical research: a narrative review. Trials 11:37
Moher D, Jadad AR, Nichol G, Penman M, Tugwell P, Walsh S (1995) Assessing the quality of randomized controlled trials. An annotated bibliography of scales and checklists. Control Clin Trials 16:62–73
Moher D, Pham B, Jones A, Cook DJ, Jadad AR, Moher M, Tugwell P, Klassen TP (1998) Does quality of reports of randomised trials affect estimates of intervention efficacy reported in meta-analyses? Lancet 352:609–613
Moher D, Cook DJ, Jadad AR, Tugwell P, Moher M, Jones A, Pham B, Klassen TP (1999) Assessing the quality of reports of randomised trials. Implications for the conduct of meta-analyses. Health Technol Assess 3:1–98
Moher D, Hopewell S, Schulz KF, Montori V, Gotzsche PC, Devereaux PJ, Elbourne D, Egger M, Altman DG (2010) CONSORT 2010 explanation and elaboration: updated guidelines for reporting parallel group randomised trials. BMJ 340:c869

National Health Service Public Health Resource Unit (2006) Critical appraisal skills programme: making sense of evidence. 10 questions to help you make sense of randomised controlled studies. Available from http://www.sph.nhs.uk/what-we-do/public-health-workforce/resources/critical-appraisals-skills-programme

Odgaard-Jensen J, Vist GE, Timmer A, Kunz R, Akl EA, Schünemann H, Briel M, Nordmann AJ, Pregno S, Oxman AD (2011) Randomisation to protect against selection bias in healthcare trials. Cochrane Database Syst Rev. doi:10.1002/14651858.MR000012.pub3

Oxman AD (1994) Checklists for review articles. BMJ 309:648–651

Piaggio G, Elbourne DR, Altman DG, Pocock SJ, Evans SJ (2006) Reporting of noninferiority and equivalence randomized trials: an extension of the CONSORT statement. JAMA 295:1152–1160

Pildal J, Hróbjartsson A, Jørgensen KJ, Hilden J, Altman DG, Gøtzsche PC (2007) Impact of allocation concealment on conclusions drawn from meta-analyses of randomized trials. Int J Epidemiol 36:847–857

Porta N, Bonet C, Cobo E (2007) Discordance between reported intention-to-treat and per protocol analyses. J Clin Epidemiol 60:663–669

Scottish Intercollegiate Guidelines Network (SIGN) (2011) SIGN 50 – a guideline developer's handbook. Available from http://www.sign.ac.uk/guidelines/fulltext/50/index.html

Thomas BH, Ciliska D, Dobbins M, Micucci S (2004) A process for systematically reviewing the literature. Providing the research evidence for public health nursing interventions. Worldviews Evid Based Nurs 1:176–184

University of Oxford (2005) Centre of Evidence Based Medicine (CEBM). Critical appraisal for therapy articles. Available from http://www.cebm.net/index.aspx?o=1157

Verhagen AP, De Vet HCW, de Bie RA, Kessels AGH, Boers M, Bouter LM, Knipschild PG (1998) The Delphi list. A criteria list for quality assessment of randomized clinical trials for conducting systematic reviews developed by Delphi consensus. J Clin Epidemiol 51:1235–1241

Verhagen AP, de Bie RA, Lenssen AF, de Vet HC, Kessels AG, Boers M, van den Brandt PA (2000) Impact of quality items on study outcome. Treatments in acute lateral ankle sprains. Int J Technol Assess Health Care 16:1136–1146

West S, King V, Carey TS, Lohr KN, McKoy N, Sutton SF, Lux L (2002) Systems to rate the strength of scientific evidence. Evid Rep Technol Assess 47:1–11

Whiting P, Rutjes AW, Dinnes J, Reitsma J, Bossuyt PM, Kleijnen J (2004) Development and validation of methods for assessing the quality of diagnostic accuracy studies. Health Technol Assess 8:1–234

Williamson PR, Gamble C (2005) Identification and impact of outcome selection bias in meta-analysis. Stat Med 24:1547–1561

Wood L, Egger M, Gluud LL, Schulz K, Jüni P, Altman DG, Gluud C, Martin RM, Wood AJG, Sterne JAC (2008) Empirical evidence of bias in treatment effect estimates in controlled trials with different interventions and outcomes: meta-epidemiological study. BMJ 336:601–605

Zaza S, Wright-De Aguero LK, Briss PA, Truman BI, Hopkins DP, Hennessy MH, Sosin DM, Anderson L, Carande-Kulis VG, Teutsch SM, Pappaioanou M (2000) Data collection instrument and procedure for systematic reviews in the guide to community preventive services. Task force on community preventive services. Am J Prev Med 18:44–74

Chapter 14
Meta-analysis I

Computational Methods

Suhail A.R. Doi and Jan J. Barendregt

Abstract Meta-analysis is now used in a wide range of disciplines, in particular epidemiology and evidence-based medicine, where the results of some meta-analyses have led to major changes in clinical practice and health care policies. Meta-analysis is applicable to collections of research that produce quantitative results, examine the same constructs and relationships, and have findings that can be configured in a comparable statistical form called an effect size (e.g. correlation coefficients, odds ratios, proportions, etc.), that is, are comparable given the question at hand. These results from several studies that address a set of related research hypotheses are then quantitatively combined using statistical methods. This chapter provides an in-depth discussion of the various statistical methods currently available, with a focus on bias adjustment in meta-analysis.

Introduction

Meta-analysis is now used in a wide range of disciplines, in particular epidemiology and evidence-based medicine where the results of some meta-analyses have led to major changes in clinical practice and health care policies. Meta-analysis is applicable to collections of research that produce quantitative results, examine the same constructs and relationships and have findings that can be configured in a comparable statistical form called an effect size (ES) (e.g. correlation coefficients, odds ratios, proportions, etc.), that is, are comparable given the question at hand. These results from several studies that address a set of related research hypotheses are then quantitatively combined using statistical methods. The set of related research hypotheses can be demonstrated at a broad level of abstraction; for example, ovarian ablation therapy for ovulation induction in polycystic ovarian syndrome

S.A.R. Doi (✉) • J.J. Barendregt
School of Population Health, University of Queensland, Brisbane, QLD, Australia
e-mail: sardoi@gmx.net

S.A.R. Doi and G.M. Williams (eds.), *Methods of Clinical Epidemiology*,
Springer Series on Epidemiology and Public Health,
DOI 10.1007/978-3-642-37131-8_14, © Springer-Verlag Berlin Heidelberg 2013

where various therapies are lumped together such as laser, wedge resection and interstitial ablation. Alternatively, it may be at a narrow level of abstraction and represent pure replications. The closer to pure replications the collection of studies is, the easier it is to argue comparability. Forms of research suitable for meta-analysis include group contrasts such as experimentally created groups, that is, comparison of outcomes between experimental and control groups and naturally or non-experimentally occurring groups (treatment, prognostic or diagnostic features). Pre-post contrasts can also be meta-analysed, for example, changes in continuous or categorical variables. Another area for meta-analysis is central tendency research such as incidence or prevalence rates and means. Association between variables can be meta-analysed, such as correlation coefficients and regression coefficients.

The meta-analysis differs from the systematic review in that the focus changes to the direction and magnitude of the effects across studies, which is what we are interested in anyway. Direction and magnitude are represented by the ES and therefore this is a key requirement for, and is what makes meta-analysis possible. It is a quantitative measure of the strength of the relationship between intervention and outcome and it encodes the selected research findings on a numeric scale. There are many different types of ES measures, each suited to different research situations. Each ES type may also have multiple methods of computation. The type of ES must be comparable across the collection of studies of interest for meta-analysis to be possible. This is sometimes accomplished through standardization when some or all of the studies use different scales (e.g. the standardized mean difference). A standard error must be calculable for that type of ES because it is needed to calculate the meta-analysis weights, called inverse variance weights (more on this later) as all analyses are weighted. Thus, it is important to abstract ES information from studies if the systematic review is to be followed up with a meta-analysis. The pooled estimate is usually computed by meta-analysis software based on the ES input selected. The software we have created is MetaXL (www.epigear.com) and it also has an option to enter the ES and standard error (SE) directly or to bypass the SE input thus allowing a multivariable adjusted ES to be entered directly.

It is therefore evident that combining quantitative data (synthesis) is what is central to the practice of meta-analysis. The basic underlying premise is that the pooled results from a group of studies can allow a more accurate estimate of an effect than an individual study because it overcomes the problem of reduced statistical power in studies with small sample sizes. However, pooling in meta-analysis must be distinguished from *simple pooling* where there is the implication that there is no difference between individual studies (or subgroups) so that it is seemingly acceptable to consider that the data from the control group of one study might just have easily come from the control group of another study. Bravata and Olkin (2001) point out that in reality, by simple pooling, we are assigning different weights to intervention and control groups and this can lead to paradoxical results. Of course, if the individual studies have the same sample size in the intervention and control groups (studies are balanced) such paradoxes will not occur and this explains why balanced designs are advocated for randomized controlled trials and why simple pooling of centres is sometimes used in such multi-centre trials. Bravata and Olkin (2001) emphasize that while *simple pooling* is obtained by combining first, then comparing, the meta-analytic method compares first, then combines. Thus, the order in which the operations of combining and comparing are carried out is the difference between simple pooling and combining data for meta-analysis and will yield different answers. Combining data via meta-analysis therefore provides a safeguard against reversals such as Simpson's paradox that can occur from simple pooling.

Common Effect Sizes

Standardized Mean Difference and Correlation

This is commonly used with group contrast research, treatment groups and naturally occurring groups where the measurements are inherently continuous. It uses the pooled standard deviation (some situations use control group standard deviation) and is called Cohen's "d" or occasionally Hedges "g". The standardized mean difference can be calculated from a variety of statistics and calculators are available for various methods and remember that any data for which you can calculate a standardized mean difference ES, you can also calculate a correlation type ES. Standardized mean difference ES has an upward bias when sample sizes are small but this can be removed with the small sample size bias transformation. If $N = n_1 + n_2$ then

$$\mathrm{ES}'_{\mathrm{sm}} = ES_{\mathrm{sm}}\left[1 - \frac{3}{4N - 9}\right]$$

Correlation has a problematic standard error formula and this is needed for the meta-analysis inverse variance weight. In this case the Fisher's Zr transformation is used:

$$\mathrm{ES}_{\mathrm{Zr}} = 0.5 \ln\left[\frac{1 + r}{1 - r}\right]$$

and results can be converted back into r with the inverse Zr transformation:

$$r = \frac{e^{2\mathrm{ES}_{\mathrm{Zr}}} - 1}{e^{2\mathrm{ES}_{\mathrm{Zr}}} + 1}$$

Odds Ratio/Relative Risk

Again this is used with group contrast research but this time there the measurements are inherently dichotomous. The odds ratio is based on a 2 by 2 contingency table and is the odds of success in the treatment group relative to the odds of success in the control group. Odds ratio/RR are asymmetric and have a complex standard error formula. Negative relationships are indicated by values between 0 and 1. Positive relationships are indicated by values between 1 and infinity. To solve this imbalance, the natural log of the odds ratio/RR is used in meta-analysis.

$$\mathrm{ES_{LOR}} = \ln[\mathrm{OR}], \quad \mathrm{ES_{LRR}} = \ln[\mathrm{RR}]$$

In this case a negative relationship is <0, no relationship = 0, and a positive relationship is >0. Results can be converted back into odds ratios/RR by the inverse natural log function.

Proportion/Diagnostic Studies

This is used in central tendency research e.g. prevalence rates and other proportions such as sensitivity and specificity. Proportions have an unstable variance and thus transformed proportions are automatically used by the software. We use the double arcsine square root transformation in MetaXL (http://www.epigear.com) and more details are given in the section below on proportions.

Pooling Effect Sizes

The Fixed Effects Model

The standard approach frequently used in weighted averaging for meta-analysis in clinical research is termed the inverse variance method or FE model based on Woolf (1955). The average ES across all studies is computed whereby the weights are equal to the inverse variance of each study's effect estimator. Larger studies and studies with less random variation are given greater weight than smaller studies. The weights (w) allocated to each of the studies are then inversely proportional to the square of the SE; thus for the ith study

$$w_i = \frac{1}{\mathrm{SE_i}^2}$$

which gives greater weight to those studies with smaller SEs.

As can be seen above, the variability within each study is used to weight each study's effect in the current approach to combining them into a weighted average as this minimizes the variance (assuming each study is estimating the same target). So, if a study reports a higher variance for its ES estimate, it would get lesser weight in the final combined estimate and vice versa. This approach, however, does not take into account the innate variability that exists between the studies arising from differences inherent to the studies such as their protocols and how well they were executed and conducted. This major limitation has been well recognized and it gave rise to the random effects (RE) model approach by DerSimonian and Laird (1986).

The Random Effects Model

A common model used to synthesize heterogeneous research is the RE model of meta-analysis. Here, a constant is generated from the homogeneity statistic Q and, using this and other study parameters, a random effects variance component (REVC) (τ^2) is generated. The inverse of the sampling variance plus this constant that represents the variability across the population effects is then used as the weight

$$w_i^* = \frac{1}{\mathrm{SE}_i^2 + \tau^2}$$

where w_i^* is the RE weight for the ith study. However, because of the limitations of the RE model, when used in a meta-analysis of badly designed studies, it will still result in biased estimates even though there is statistical adjustment for ES heterogeneity (Senn 2007). Furthermore, such adjustments, based on an artificially inflated variance, lead to a widened confidence interval, supposedly to reflect ES uncertainty, but Senn (2007) has pointed out that they do not have much clinical relevance.

The weight that is applied in this process of weighted averaging with an RE meta-analysis is achieved in two steps:

- Step 1: Inverse variance weighting
- Step 2: Un-weighting of this inverse variance weighting by applying an REVC that is simply derived from the extent of variability of the ESs of the underlying studies.

This means that the greater this variability in ESs (otherwise known as heterogeneity), the greater the un-weighting and this can reach a point when the RE meta-analysis result becomes simply the un-weighted average ES across the studies. At the other extreme, when all ESs are similar (or variability does not exceed sampling error), no REVC is applied and the RE meta-analysis defaults to simply a fixed effect meta-analysis (only inverse variance weighting). Al Khalaf et al. (2011) have pointed out that the extent of this reversal is solely dependent on two factors:

1. Heterogeneity of precision
2. Heterogeneity of ES

Since there is absolutely no reason to automatically assume that a larger variability in study sizes or ESs automatically indicates a faulty larger study or more reliable smaller studies, the re-distribution of weights under this model bears no relationship to what these studies have to offer. Indeed, there is no reason why the results of a meta-analysis should be associated with this method of reversal of the inverse variance weighting process of the included studies. As such, the changes in weight introduced by this model (to each study) results in a pooled estimate that can have no possible interpretation and, thus, bears no relationship with what the studies actually have to offer.

To compound the problem further, some statisticians are proposing that we take an estimate that has no meaning and compute a prediction interval around it. This is akin to taking a random guess at the effectiveness of a therapy and under the false belief that it is meaningful try to expand on its interpretation. Unfortunately, there is no statistical manipulation that can replace commonsense. While heterogeneity might be due to underlying true differences in study effects, it is more than likely that such differences are brought about by systematic error. The best we can do in terms of addressing heterogeneity is to look up the list of studies and attempt to un-weight (from inverse variance) based on differences in evidence of bias rather than ES or precision that are consequences of these failures.

Problems with These Conventional Models

One problem with meta-analysis is that differences between trials, such as sources of bias, are not addressed appropriately by current meta-analysis models. Bailey (1987) lists several reasons for such differences: chance, different definitions of treatment effects, credibility-related heterogeneity (quality), and unexplainable and real differences. An important explainable difference is credibility-related heterogeneity (quality) and this has been defined by Verhagen et al. (2001) as the likelihood of the trial design generating unbiased results that are sufficiently precise to allow application in clinical practice. The flaws in the design of individual studies have obvious relevance to creating heterogeneity between trials as well as an influence on the magnitude of the meta-analysis results. If the quality of the primary material is inadequate, this may falsify the conclusions of the review, regardless of the presence or absence of ES heterogeneity. The need to address heterogeneity in trials via study-specific assessment has been obvious for a long time and the solution involves more than just inserting a random term based on ES heterogeneity as is done with the RE model.

Previous studies that have attempted to investigate incorporation of some study-specific components in the weighting of the overall estimate concluded that incorporating such information into weights provided inconsistent adjustment of the estimates of the treatment effect. Although these authors follow the same assumption as we do that studies with deficiencies are less informative and should have less influence on overall outcomes, methodology was flawed and such attempts therefore did not reduce bias in the pooled estimate, and may have resulted in an increase in bias.

A study score-adjusted model that overcomes several limitations has been introduced by Doi and Thalib (2008, 2009). The rationale was that in a group of homogeneous trials, it is assumed that because the ESs are homogeneous, the studies are all estimating the same target effect (we can call this a type A trial). In this situation, the inverse variance weights of Woolf (1955) will minimize the variance since the mean squared error (MSE) = expected(estimate − true)^2 = variance + (bias)^2. Bias is zero if the underlying true ESs are equal and thus minimizing variance is optimal and the weighted MSE = variance. It is thought that the inverse variance-weighted analysis tests the null hypothesis that all studies in the meta-analysis are identical and show no effect of the intervention under consideration regardless of homogeneity. This requires the assumption that trials are exchangeable so that if one large trial is null and multiple small trials show an effect, the large trial essentially decreases evidence against the null hypothesis. Exchangeability, however, is likely to be conditional only, as discussed later, and thus this is a big assumption. Therefore, if we do not believe the trials are exchangeable then, in this situation, we have two alternatives: either the trials have been affected by bias even though the underlying true effects are identical (we can call these type B trials) or the trials represent different underlying true effects (we can call these type C trials). In the former case, the trial ES from a biased trial might seem like it is coming from a different underlying true effect, thus giving the impression that the trials represent different underlying true effects. In type B trials, inverse variance weights do not minimize the variance, it just exaggerates it and creates gross bias in these situations. Furthermore, any set of weights in a type A situation estimates the same target, but in a type B situation each set of weights estimates a different target. Thus, inverse variance weights in the latter situation just increase bias and are not optimal for type B trials. Thus, in type B trials, we would want to use situation-specific weights.

One such situation-specific weight that has been suggested for type B trials is weighting according to the probability (Q_i) of credibility (internal validity or quality) of the studies making up the meta-analysis. Although this can correct for distortions due to systematic error, it can also introduce errors of another type. For example, a study of a small sample that is not representative of the underlying population may get a large quality weighting and this can skew the data. It might thus be informative to weight according to precision and then redistribute the weights according to situation-specific requirements. In this case, the importance of smaller good quality studies are upgraded only if the larger or more precise studies are deemed poor by its situation-specific weight.

This line of thought is not new as this is precisely what the RE model attempts to do. The unfortunate thing, however, is that the situation-specific weight used in this particular model is an index of the variability of the ESs across trials and the same situation-specific weight is applied to all trials (the RE model). It becomes quite clear that the type B meta-analysis differ from the RE model in that between-study variability is visualized as a fixed rather than a random effect and thus represents an extension of a fixed effects model that can address heterogeneity. In type B trials, the expectation is that the expected value of the study estimate differs from the grand (real) mean (μ) by an amount β_j and the true (study-specific) mean (θ_j) for study j is given by $\theta_j = \mu + \beta_j$. The divergence, however, is that the β_j are not interpreted as a random effect with type B studies and thus do not have a common variance. The philosophy behind the random effect construct is that it presupposes that the study effects are randomly sampled from a population with a varying ($\sigma_\tau^2 > 0$) underlying parameter of interest. Overton (1998) thus has stated that if the studies included in the meta-analysis differ in some systematic way from the possible range in the population (as is often the case in the real world), they are not representative of the population and the RE model does not apply, at least according to a strict view of randomization in statistical inference.

In addition, with the RE model, the weight of the larger studies are redistributed to smaller studies but τ^2 has a decreasing effect as study precision declines. The size of τ^2 is determined by how heterogeneous the ESs are and if τ^2 is zero, the RE model defaults to the FE model. If we focus on the largest study, the bigger its difference from other studies, the bigger the τ^2 and the decrease in weight of this study. Al Khalaf et al. (2011) demonstrates that τ^2 has a U-shaped association with ES in the largest study, being minimal when the largest study conforms to other study ESs, and as this ES departs from that of other studies, τ^2 increases. The weight of the largest study then declines as τ^2 increases. However, while the biggest individual study weight decrements associated with bigger τ^2 follow a predictable pattern, the impact of different τ^2 values on the pooled estimate is unpredictable. This happens because, although individual study weight changes are predictable from τ^2, the relationship of weight gain across smaller studies bears no relationship to which study shows the most ES heterogeneity, or indeed any tangible information from the study.

The Quality Effects Model

In order to rectify this situation, an alternative approach was proposed by Doi and Thalib (2008) and subsequently modified in 2011 and 2012. The main reasoning was: suppose there are K studies in a set of studies that belong to a meta-analysis and x_j and w_j are random variables representing the ES and normalized (sum to 1) weights, respectively, with the study labels $j = 1,\ldots,K$. The expected value of x_j was taken to be the underlying parameter (μ) being estimated. However, in this situation, the ESs are assumed to be similar in the sense that the study labels

($j = 1,\ldots,K$) convey no information and are thus considered independent and identically distributed (IID). The reality is that each of these labels (representing independent studies) is associated with specific information about the likelihood of systematic bias (β_j) and thus for all j the x_j are in fact only conditionally IID and would be estimating a specific biased parameter. Assuming that heterogeneity derives from essentially non-random systematic error and randomness is only obtained via a random permutation of the indices $1,2,\ldots,K$, then details about the design of study j do provide information about these systematic errors and can be represented by a hierarchical model for each study:

$$\beta_j \sim N(\beta, \phi^2) \quad \text{(bias effects)}$$

$$(x_j \mid \widehat{\mu + \beta_j}) \overset{\text{indep}}{\sim} N(\widehat{\mu + \beta_j}, {\sigma_j}^2 + \phi_j^2) \quad \text{(study)}$$

The bottom level of underlying effects, the study level of the hierarchical model, says that because of relevant differences in methodology and systematic errors, each study has its own underlying treatment effect $\mu + \beta_j$, and the observed ES differences x_j are like random draws from a normal distribution with mean $\widehat{\mu + \beta_j}$ and variance ${\sigma_j}^2 + \phi_j^2$ (the normality is reasonable because of the central limit theorem). Thus, a suitable linear model for the jth study (not considering across all studies) can be written as

$$x_j = \widehat{\mu + \beta_j} + \varepsilon_j \tag{14.1}$$

and for each study

$$\mathrm{E}(\mathrm{x}_j) = \widehat{\mu + \beta_j}$$

Also, under the assumption of no prior information about weights (w_j) except that they sum to 1, they will be equally distributed with the expected value of w_j being $1/K$ for all j. If $c = \mathrm{Cov}(\mathrm{w}_j, \mathrm{x}_j)$ is the covariance of these random quantities across all studies, then

$$\mathrm{E}(\mathrm{w}_j\mathrm{x}_j) = \mathrm{Cov}(\mathrm{w}_j, \mathrm{x}_j) + \mathrm{E}(\mathrm{x}_j)\mathrm{E}(\mathrm{w}_j) = c + \widehat{\mu + \beta_j}/K$$

and thus summing across all studies,

$$\mathrm{E}\left(\sum_{j=1}^{\mathrm{K}} (\mathrm{w}_j\mathrm{x}_j)\right) = \mu + \frac{1}{K}\sum \beta_j + Kc$$

since $\mathrm{E}(w_j) = 1/K$.

Thus, it is clear that if we use empirical weights, c *is* not zero, $\Sigma\beta_j$ is also not zero and the meta-analytic estimate for μ is biased. It is probably true, as suggested by Shuster (2010), that the unweighted estimate is a less biased estimate in situations where w_j and x_j are correlated. However, it is clear that an unbiased estimate of μ will not be provided unless the average $\beta_j = 0$, so systematic error also leads to increase in bias.

Everything hinges on the variance and, therefore, the mere observation that the unweighted estimate is likely to be unbiased does little to reaffirm our confidence in its utility without a simultaneous measure of its global error (with respect to its parameter). The MSE thus has to be minimized and the fact that bias is included as a component is important because the judgment of the performance of the model depends on the trade-off between the amount of bias and the variability.

It may be noted that for a particular study,

$$\mathrm{Var}(\mathrm{x}_j\mathrm{w}_j) = (\sigma_j^2 + \phi_j^2)\mathrm{w}_j^2$$

Therefore,

$$\mathrm{Var}\left(\sum \mathrm{x}_j\mathrm{w}_j\right) = \sum (\sigma_j^2 + \phi_j^2)\mathrm{w}_j^2 \qquad (14.2)$$

Also, under the constraint that $\Sigma w = 1$ and only if ${\sigma_j}^2 + \phi_j^2$ was equal for all K studies, does the variance attain its minimum value for equal weights, and its maximum when all weights except one are zero. This is not the case from Eq. 14.2 and the naturally weighted average is expected to have a poor bias–variance trade-off. The only logical solution therefore is to discount studies that are expected to have an inflated value for β_j. This can be achieved by linking β_j to the probability that a study is credible as follows. If $\beta = 0$ and if

$$\sum_{j=1}^{K} \beta_j^2/K = \phi^2$$

then

$$Q_j = \phi^2/(\phi^2 + \phi_j^2)$$

which can be interpreted as the probability that study j is credible as described previously by Spiegelhalter and Best (2003) or Turner et al. (2009). Therefore,

$$\phi_j^2 = (\phi^2 - Q_j\phi^2)/Q_j$$

What this means is that as Q_j and the individual study bias variance (ϕ_j^2) are inversely related and thus an inverse discounting system for such studies based on Q_j should be optimal if the expected increase in bias ends up being traded off by larger decreases in variance. This is a logical conclusion also reiterated by Burton et al. (2006) as any method that results in an unbiased estimate but has large variability cannot be considered to be have much practical use.

To discount by quality requires computation of an adjusted Q_j first as follows (See Doi et al. 2011, 2012):

$$Q_j(\text{adj}) = \begin{cases} \left(\dfrac{\left(\sum\limits_{j=1}^{K} Q_j\right)\tau_j}{\left(\sum\limits_{j=1}^{K} \tau_j\right)(K-1)} \right) + Q_j & \text{if } (\exists \mathrm{Q}_j)\ \mathrm{Q}_j < 1 \\ Q_j & \text{otherwise.} \end{cases}$$

where

$$\tau_j = \frac{iw_j - (iw_j \times Q_j)}{K-1}$$

and iw_j is the inverse variance weight of study j, Q_j is the credibility of study j ranging from 0 to 1 and K is the number of studies in the meta-analysis. From the adjusted quality parameter, a quality adjustor is then computed given by

$$\hat{\tau}_j = \left(\left(\sum_{j=1}^{K} \tau_j\right) K \frac{Q_j(\text{adj})}{\sum\limits_{j=1}^{K} Q_j(\text{adj})} \right) - \tau_j$$

This is then used to compute the study bias-specific variance component $\hat{Q}_j$ as follows:

$$\hat{Q}_j = Q_j + \left(\frac{\hat{\tau}_j}{iw_j}\right)$$

What these equations do is replace the REVC with study-specific variance components so that the target this meta-analysis is estimating becomes meaningful. Given that the final weight for the study is $w_j^{\delta} = iw_j \hat{Q}_j$, the final summary estimate is then given by

$$\bar{x}_{\text{QE}} = \frac{\sum (w_j^{\delta} \times x_j)}{\sum w_j^{\delta}} = \frac{\sum (\hat{Q}_j \times iw_j \times x_j)}{\sum (\hat{Q}_j \times iw_j)}$$

where $\bar{x}$ is the pooled ES measure and it has a variance (V) given by

$$V_{\text{QE}} = \sum {\sigma^2}_j \left(\frac{w_j^{\delta}}{\sum w_j^{\delta}} \right)^2$$

Given that $iw_j = 1/\sigma_j^2$, this reduces to:

$$V_{\text{QE}} = \frac{\sum (\hat{Q}_j^{\,2} \times iw_j)}{\left(\sum (\hat{Q}_j \times iw_j) \right)^2}$$

However, there is expected to be significant overdispersion and thus this variance estimate underestimates the true variance and can lead to a confidence interval with poor coverage. To rectify this, a correction factor (CF) has been proposed for overdispersion based on iterative simulation studies using the Q statistic (χ_c) as follows (Doi et al. 2011):

$$\text{CF} = \left(1 - \max\left[0, \frac{\chi_c - (\text{K} - 1)}{\chi_c} \right] \right)^{0.25}$$

For computation of the variance of the weighted average, the variance of each study is then inflated to the power CF as follows:

$$iw'_j = \frac{1}{\left(\sigma_j^2\right)^{\text{CF}}} \quad \text{if } \sigma_j^2 < 1 \quad \text{or} \quad iw'_j = \frac{1}{\left(\sigma_j^2\right)^{(2-\text{CF})}} \quad \text{if } \sigma_j^2 \geq 1$$

This can then be used to update V_{QE} as follows:

$$V_{\text{QE}} = \frac{\sum (\hat{Q}_j^{\,2} \times iw'_j)}{\left(\sum (\hat{Q}_j \times iw'_j) \right)^2}$$

Assuming the distribution of these estimates are asymptotically normal, the 95 % confidence limits are easily obtained by

$$95\,\%\ \text{CI} = \overline{\text{ES}} \pm 1.96(\sqrt{\text{V}_{\text{QE}}})$$

It becomes quite clear, that the quality-based method differs from the RE model in that between-study variability is visualized as a fixed rather than a random effect and thus represents an extension of a fixed effects model that can address heterogeneity. In both the classic random effect method and the quality-based method, the β_j is taken to be the difference between the grand (real) mean (μ) and the true (study-specific) mean (x_j) for study j ($\beta_j = x_j - \mu$). The divergence, however, is that the β_j are not interpreted as a random effect with the quality-based method and thus do not have a common variance. The philosophy behind the random effect construct is that it presupposes that the x_j values are randomly sampled from a population with a varying underlying parameter of interest ($\tau^2 > 0$). However, if the studies included in the meta-analysis differ in some systematic way from the possible range in the population (as is often the case in the real world), they are not representative of the population and the RE model does not apply, at least according to a strict view of randomization in statistical inference (Overton 1998). The quality-based method therefore corrects this by interpreting the β_j as a fixed effect related to the study itself (based on systematic or related errors) and thus the effect of a varying target created by this bias can be minimized by discounting studies where within-study bias variance (ϕ_j^2) is likely to be large relative to between-study bias variance (ϕ^2). Such discounting requires a robust mechanism to avoid increasing bias and to simultaneously allow incorporation of sampling errors into the model as detailed above based on previous work on this subject. As mentioned by Eisenhart (1947), which situation applies to the model is the deciding factor in determining whether effects are to be considered as fixed or random and when inferences are going to be confined to the effects in the model, the effects are considered fixed.

While with this model we assume that non-credibility leads to bias in the ES, this supposition is backed by clear evidence from several authors such as Balk et al. (2002), Conn and Rantz (2003), Egger et al. (2003), Moher et al. (1998), Schulz et al. (1995) and others suggesting that inadequate methodology correlates with bias in the estimation of treatment effects. However, there could be instances where lack of credibility does not lead to bias in the estimation of treatment effects (or alternatively where such biases may have been obscured by the lack of credibility). In such cases, the quality effects (QE) model is still valid and credibility information results simply in decreased confidence (wider confidence intervals) in the pooled estimate. We do not delete lower quality studies because every study has something to add to the weighted estimate. We do not know what the relationship of study-specific scores are to the magnitude or direction of bias. However, if this weighting is not based on study- or goal-specific attributes, then the weighted estimate loses meaning. A sensitivity analysis, on the other hand, can only tell us that subgroups are heterogeneous but not what the true estimate is likely to be. In studies that vary due to systematic error, study-specific scores can lead to the best approximation of the true ES. The letter would not be possible with either the RE model or sensitivity analyses.

When weighting study estimates by their study-specific scores, we must keep in mind that these scores do not tell us the direction or magnitude of the change in ES that is attributable to that score. The QE method of Doi and Thalib (2008), is not constrained by this limitation, because, unlike previous methods, it does not adjust a study weight directly but redistributors it in relation to all other study weights based on its quality status. This is exactly what the RE model does too, the major difference being that the latter adds on weight to smaller studies without any rationale for doing so and the process ultimately becomes random. This is because τ^2 is not individualized to each study as $\hat{\tau}_i$ is in the QE model. A gradual increase in weight of smaller studies with quality is seen but not with ES heterogeneity. This also explains why previous attempts by Berard and Bravo (1998) or Tritchler (1999) to incorporate study-specific scores into weights have failed to provide sufficient adjustment of the estimates of treatment effects as they failed to consider ramdom error or counterintuitively decided to incorporate study-specific scores over the random redistribution in an RE model.

Greenland (1994) suggested more than a decade ago that quality scoring merges objective information with arbitrary judgments in a manner that can obscure important sources of heterogeneity among study results. He gave the example of dietary quality scoring in the Nurses Health Study and states that the result would likely indicate no diet effects associated with disease if the effects of important quality items are confounded within strata of the summary quality score. The problem is to use the information regarding quality in this way. If we viewed the diet quality score as the probability that a nurse's diet is accurately measured, we would be able to rank nurses from best to worst reliability of dietary information. Even if this ranking is subjective or poor, we would still be more confident about the relationship between diet and disease in high scorers than in low scorers. This is the correct use of quality scores, but cannot be demonstrated with conventional meta-analysis models (Al Khalaf et al. 2011) given that the spread of precision and ES take precedence over stratification by quality score. The fact that previous authors used scores as exclusion criteria or to sequentially combine trial results using these models would only increase bias by altering the range of precision and ES differences among stratified studies. This is probably the reason why many authors such as Balk et al. (2002), Herbison et al. (2006), Juni et al. (1999) and Whiting et al. (2005) all report that stratification of meta-analyses by quality score has no clear impact on the pooled estimate.

Study-specific assessment has not, until now, found an acceptable means of becoming an important part of meta-analyses. More than half of published meta-analyses do not specify in the methods whether and how they would use study-specific assessment in the analysis and interpretation of results, and only about 1 in 1,000 systematic reviews consider weighting by quality score (Moja et al. 2005). This is probably because of the lack, until now, of an adequate model to do so and therefore those meta-analyses that had an a priori conceptualization of quality simply linked it to the interpretation of results or to limit the scope of the review. Although there is no gold standard and we still do not know how best to measure quality, this is not an obstacle to QE analysis because it works with any quality score. Given that we have demonstrated that the RE model randomly adjusts estimates of treatment effects in a meaningless fashion, it may now be time to switch from observed random statistical ES heterogeneity to models that are based on measured study-specific estimates of their heterogeneity.

The Special Case with Proportions in Meta-analysis

Just about all epidemiologists habitually speak of the prevalence rate, but prevalence is defined as a proportion: the number of cases in a population divided by the population number. This definition implies that (1) prevalence is always between 0 and 1 (inclusive), and (2) the sum over categories always equals 1.

The definition of prevalence is the same as the definition of the binomial distribution (number of successes in a sample), and therefore the standard assumption is that prevalence follows a binomial distribution. With the main meta-analysis methods based on the inverse variance method (or modifications thereof), the binomial equation for variance (expressed as a proportion) can be used to obtain the individual study weights:

$$\mathrm{Var}(p) = \frac{p(1-p)}{N}$$

where p is the prevalence proportion and N is the population size.

With the variance of the individual studies nailed down, the pooled prevalence estimate P then becomes (according to the inverse variance method)

$$P = \frac{\sum_i \frac{p_i}{\mathrm{Var}(p_i)}}{\sum_i \frac{1}{\mathrm{Var}(p_i)}}$$

with SE

$$\mathrm{SE}(P) = \sqrt{\sum_{\mathrm{i}} \frac{1}{\mathrm{Var}(p_i)}} \tag{14.3}$$

The confidence interval of the pooled prevalence can then be obtained by

$$\mathrm{CI}_\gamma(P) = P \pm Z_{\alpha/2}\mathrm{SE}(P)$$

where $Z_{\alpha/2}$ denotes the appropriate factor from the standard normal distribution for the desired confidence percentage (e.g. $Z_{0.025} = 1.96$).

While this works fine for prevalence proportions around 0.5, increasing problems arise when the proportions get closer to the limits of the 0...1 range. The first problem is mostly cosmetic: the equation for the confidence interval does not preclude confidence limits outside the 0...1 range. While this is annoying, the second problem is much more substantial: when the proportion becomes small or big, the variance of the study is squeezed towards 0 (see Eq. 14.3). As a consequence, in the inverse variance method, the study gets a large weight. A meta-analysis of prevalence according to the method described above therefore puts undue weight on the studies at the extreme of the 0...1 range.

One way to avoid the problem of variance instability with extremes of prevalence is to estimate the SE, not using the individual proportions, but the overall proportion:

$$\mathrm{Var}(p_{ic}) = \frac{p_{\mathrm{total}}(1 - p_{\mathrm{total}})}{N_{ic}}$$

The numerator is now the same for every study and there is no longer the problem where studies with proportions near 50 % get much smaller weights than studies with proportions much smaller or much larger than 50 %. This approach also avoids the problem where a study has 100 % prevalence proportion.

$$w_{ic} = \frac{N_{ic}}{p_{\mathrm{Ctotal}}(1 - p_{\mathrm{Ctotal}})}$$

where $c = 1,\ldots,k$ denotes a particular category out of k categories. In a fixed effect model, use of the pooled proportion to get individual variances would be exactly the same as using individual proportions for variances because the SE of the pooled prevalence in category c becomes

$$\frac{1}{\sum w_{ic}} = \frac{p_{\mathrm{Ctotal}}(1 - p_{\mathrm{Ctotal}})}{\sum N_{ic}}$$

Since each study gets the same weight across categories, this method ensures that the pooled category prevalences sum to 1. However, the confidence interval does not preclude confidence limits outside the 0...1 range, so that problem persists.

The Logit Transformation

To address this issue of estimates falling outside the 0...1 range, the logit transformation was proposed and, at that time, it was thought that it would address both the problems mentioned above. It is given by

$$\mathrm{logit}(p) = \ln\left(\frac{p}{1 - p}\right)$$

with variance

$$\mathrm{Var}(\mathrm{logit}(p)) = \frac{1}{Np} + \frac{1}{N(1-p)}$$

The logit of a proportion has an approximately normal distribution, and as it is unconstrained, it was thought it would avoid the squeezing of the variance effect. The meta-analysis is then carried out on the logit transformed proportions, using the inverse of the variance of the logit as the study weight. For the final presentation, the pooled logit and its confidence interval are back transformed to a proportion using

$$P = \frac{\exp(\mathrm{logit}(P))}{\exp(\mathrm{logit}(P)) + 1}$$

While the logit transformation solves the problem of estimates falling outside the 0...1 limits, unfortunately, it does not succeed in stabilizing the variance; rather there is a reversal of the variance instability of the non-transformed proportions and studies with proportions close to 0 or 1 get their variance estimates grossly magnified and vice versa for proportions around 0.5. The variance instability that plagued non-transformed proportions thus persists even after logit transformation. It has therefore been suggested that, as a rule of thumb, the logit transformation should be used when prevalences are less than 0.2 or more than 0.8.

The Freeman–Tukey Variant of the Double Arcsine Square Root Transformation

The Freeman–Tukey transformation addresses both the problems mentioned above. It is given by

$$t = \sin^{-1}\sqrt{\frac{x_i}{n_i + 1}} + \sin^{-1}\sqrt{\frac{x_i + 1}{n_i + 1}}$$

with variance

$$\mathrm{Var}(t) = \frac{1}{n_i + 0.5}$$

The Freeman–Tukey transformed proportion has an approximately normal distribution, and, by being unconstrained, avoids the squeezing of the variance effect. A meta-analysis can be carried out on the transformed proportions, using the inverse of the variance of the transformed proportion as the study weight. For final presentation, the pooled Freeman–Tukey transformed proportion and its confidence interval are back transformed to a proportion using

$$\bar{P}(\bar{t}) = \begin{cases} 0.5\{1 - \text{sgn}(\cos\bar{t})[1 - (\sin\bar{t} + (\sin\bar{t} - 1/\sin\bar{t})/[1/\bar{v}])^2]^{0.5}\} & \text{if } p/\hat{v} \geq 2 \\ [\sin(\bar{t}/2)]^2 & \text{otherwise.} \end{cases}$$

where $\bar{P}$ is the pooled prevalence, $\bar{v}$ is the pooled variance and $\bar{t}$ is the pooled t.

The lower (LCL) and upper (UCL) confidence limits of the pooled prevalence are given by

$$\text{LCL} = \begin{cases} 0.5\left\{1 - \text{sgn}(\cos\bar{t})\left[1 - \left(\sin\bar{t} + (\sin\bar{t} - 1/\sin\bar{t})/\left[1/\hat{v}\right]\right)^2\right]^{0.5}\right\} & \text{if } \bar{p}/\hat{v} \geq 2 \\ 0 & \text{otherwise.} \end{cases}$$

$$\text{UCL} = \begin{cases} 0.5\left\{1 - \text{sgn}(\cos\bar{t})\left[1 - \left(\sin\bar{t} + (\sin\bar{t} - 1/\sin\bar{t})/\left[1/\hat{v}\right]\right)^2\right]^{0.5}\right\} & \text{if } (1-\bar{p})/\hat{v} \geq 2 \\ 1 & \text{otherwise.} \end{cases}$$

Multi-category Prevalence

The discussion so far has implicitly been about two categories (disease present or absent). But in some instances k-category prevalences may be meta-analysed where $k > 2$ (e.g. mild, moderate and severe disease), and this complicates matters.

Using the previously mentioned non-transformed and logit transformed proportions, it would not be possible to meta-analyse each category separately, since the variance of both p and logit(p) depends on p itself; this implies that the same study could get a different weight in each category, which seems hard to justify. Moreover, the sum over the pooled category prevalences would not add up to 1; another drawback.

To correct this problem, we can use the double arcsine square root transformed proportion where the SE is no longer dependent on the size of the proportion, so that both equal weights across categories and confidence limits within the 0...1 range are achieved without the need for overall proportions.

However, we again have a problem with the RE and QE models in that we need a common study estimate for Cochran's Q that can be used in weighting with the RE model and overdispersion correction with the QE model if we are to interpret the pooled proportions as dependent proportions that add to 1. This was not too difficult to conceptualize because if we believe that the ES variations across studies in one category of proportions is not independent of variations in the other categories, the maximum category Q value would be the best and most conservative estimate of a common study Q that can be applied to categorical prevalences that would allow pooled prevalences to be considered dependent and thus sum to 1. Thus, with three or more categories, the actual study heterogeneity Q value can be determined by the category with the most heterogeneity.

The only minor drawback is that pooled prevalences do not add exactly to 1 across categories when back transformed to the actual proportion because the non-linear nature of the double arcsine transformation causes the sum over the back-transformed category prevalences to become unequal to 1 (since the transformed proportion (t) can have several values (albeit close together) for the same value of prevalence). Thus, while the sum of back-transformed pooled proportion comes close to 1, it would still not exactly add to 1, unlike the standard prevalences. The error is small and thus can be corrected simply by adjusting the pooled prevalence ($\hat{P}$) in each category after pooling and back transformation:

$$\text{Adjusted } \hat{P}_c = \frac{\hat{P}_c}{\sum_{c=1}^{k} \hat{P}_c}$$

This is then the final prevalence in each category. The confidence intervals however need no adjustment. This procedure is available in MetaXL (www.epigear.com).

Prevalence Studies from Different Populations

One further consideration is type C trials, which usually deal with the burden of disease where true differences across populations are expected. A study of, for example, 1,000 respondents is equally useful for examining the mortality in a country with ten million inhabitants as it would be in a country with a population of only one million. Without weighting, any figures that combine data for two or more countries would overrepresent smaller countries at the expense of larger ones. So a population size weight is needed to make an adjustment to ensure that each country risk is represented in the pooled estimate proportional to its population size. Although such weighting has been attempted previously by Batham et al. (2009), it has been improperly applied. The best method is to assign a proportional weight between 0 and 1 for each study in relation to the largest based on the underlying population size. The population size weight (P(weight)) is thus the proportional weight $P(\text{size})_i/P(\text{size})_{\max}$. However, we must emphasize that inverse variance weights have no rule here and this may more appropriately be considered "risk adjustment" or standardization rather than meta analysis (see Appendix 2).

Appendix 1: Need for an Overdispersion Correction

In a study with overdispersed data, the mean or expectation structure (θ) is adequate but the variance structure [$\sigma^2(\theta)$] is inadequate. Individuals in the study can have the outcome with some degree of dependence on study-specific parameters unrelated to the intervention. If such data are analysed as if the outcomes were independent, then sampling variances tend to be too small, giving a false sense of precision. One approach is to think of the true variance structure as following the form [$\varphi(\theta)\sigma^2(\theta)$]; however, it is complex to fit such a form. As a simpler approach, we suppose $\varphi(\theta) = c$, so that the true variance structure [$c\sigma^2(\theta)$] is some constant multiplier of the theoretical variance structure. A common method of estimating c suggested used by Lindsey (1999) or Tjur (1998) is to use the observed chi-squared goodness of fit statistic for the pooled studies divided by its degrees of freedom:

$$c = \chi^2/\text{df}$$

If there is no overdispersion or lack of fit, $c = 1$ (because the expected value of the chi-squared statistic is equal to its degrees of freedom) and if there is, then $c > 1$. In a meta-analysis, this goodness of fit chi-squared divided by its df is equal to H^2 as defined by Higgins and Thompson (2002).

The problem of using the overdispersion parameter as a constant multiplier of the variances of each study in the meta-analysis presupposes that, for a constant increase in this parameter, there is a constant increase in variance. This means that the impact of the parameter is not capped and a point is eventually reached where there is overinflation of the variances for a given level of overdispersion resulting in overcorrection and confidence intervals that are too wide. In order to reduce the impact of large values of H^2, we can transform H^2 to its reciprocal and use this to proportionally inflate the variances. Higgins and Thompson (2002) also defined an I^2 parameter, which is an index of dispersion that is restricted between 0 (no dispersion) and 1. If we reverse the I^2 scale (by subtracting it from 1) so that no dispersion (only sampling error) is now 1 as opposed to 0, then $(1 - I^2)$ is indeed the reciprocal of H^2. We thus used $(1 - I^2)$ as an exponent to proportionally inflate study variances < 1. For variance > 1, we used 2 minus this overdispersion parameter (which reduces to $[I^2 + 1]$) as the inflation factor. Additional rescaling was done by scaling $(1 - I^2)$ to various roots and using the simulation described above to see the impact on coverage of the confidence interval. The fourth root was found to result in an acceptable simulated coverage of the confidence interval around 95 %. We thus used $[(1 - I^2)^{1/4}]$ as the final overdispersion correction factor. This is also equivalent to $(1/H^2)^{1/4}$. This correction was then used to inflate the variances of individual studies resulting in a more conservative meta-analysis pooled variance. Even if the accuracy of this approximation is questionable, common sense suggests that it is better to perform this correction, implicitly making the (more or less incorrect) assumption that the distribution of c is approximated well enough by a χ^2 distribution with $k - 1$ degrees of freedom than not to perform any correction at all, implicitly making the (certainly incorrect) assumption that there is no overdispersion in the data (Tjur 1998). This adjustment in the QE model corrects for overdispersion within studies that affect the precision of the pooled estimate, not for heterogeneity between studies that affect the estimate itself.

Appendix 2: Quality Scores and Population Impact Scores

For a QE type of meta-analysis, a reproducible and effective scheme of quality assessment is required. However, any quality score can be used with the method and thus we are not constrained to any one method. There are many different quality assessment instruments and most have parameters that allow us to assess the likelihood for bias. Although the importance of such quality assessment of experimental studies is well established, quality assessment of other study designs in systematic reviews is far less well developed. The feasibility of creating one quality checklist to apply to various study designs has been explored by Downs and Black (1998), and research has gone into developing instruments to measure the methodological quality of observational studies in meta-analyses (see Chap. 13). Nevertheless, there is as yet no consensus on how to synthesize information about quality from a range of study designs within a systematic review, although many quality assessment schemes exist. Concato (2004) suggests that a more balanced view of observational and experimental evidence is necessary. The way Q_i is computed from the score for each study and the additional use of population weights (for burden of disease or type C studies) is depicted in Table 14.1. The population weights are applied as a method of standardization of the group pooled estimates where there is a single estimate per group. The population weighted analysis does not use inverse variance weighting and if a rate is being pooled would give an equivalent result to direct standardization used in epidemiology. Rates have a problematic variance but can be based on a normal approximation to the Poisson distribution:

$$\mathrm{Var}_{\mathrm{rate}} = O \times \left(\frac{K}{P}\right)^2$$

where O are the observed events, P is the person-time of observation and K is a constant multiplier. In the computation, zero rates can be imputed to have variances based on a single observed event as a continuity correction.

Table 14.1 Hypothetical calculation of Q_i for use in QE meta-analyses[a]

Study name	Points assigned based on a quality checklist (maximum possible, e.g. 12 points)	Probability that study is credible (Q_i)	Pooled estimate by country using Q_i	Population at risk (if applicable and only for burden of disease studies)	Population impact score (normalized)	Population weighted estimate (equivalent to direct standardization) using the population weights but not including inverse variance weights
Country 1						
Study A	5	5/12 = 0.42	Estimate for country 1	400,000	400,000/400,000 = 1	Standardized estimate
Study B	7	7/12 = 0.58				
Study C	10	10/12 = 0.83				
Country 2						
Study D	5	5/12 = 0.42	Estimate for country 2	100,000	100,000/400,000 = 0.25	
Study E	7	7/12 = 0.58				
Study F	10	10/12 = 0.83				

[a]The four columns on the right apply only in burden of disease (type C) studies

Bibliography

Al Khalaf MM, Thalib L, Doi SA (2011) Combining heterogenous studies using the random-effects model is a mistake and leads to inconclusive meta-analyses. J Clin Epidemiol 64:119–123
Bailey KR (1987) Inter-study differences: how should they influence the interpretation and analysis of results? Stat Med 6:351–360
Balk EM, Bonis PA, Moskowitz H, Schmid CH, Ioannidis JP, Wang C, Lau J (2002) Correlation of quality measures with estimates of treatment effect in meta-analyses of randomized controlled trials. JAMA 287:2973–2982
Batham A, Gupta MA, Rastogi P, Garg S, Sreenivas V, Puliyel JM (2009) Calculating prevalence of hepatitis B in India: using population weights to look for publication bias in conventional meta-analysis. Indian J Pediatr 76:1247–1257
Berard A, Bravo G (1998) Combining studies using effect sizes and quality scores: application to bone loss in postmenopausal women. J Clin Epidemiol 51:801–807
Bravata DM, Olkin I (2001) Simple pooling versus combining in meta-analysis. Eval Health Prof 24:218–230
Burton A, Altman DG, Royston P, Holder RL (2006) The design of simulation studies in medical statistics. Stat Med 25:4279–4292
Concato J (2004) Observational versus experimental studies: what's the evidence for a hierarchy? NeuroRx 1:341–347
Conn VS, Rantz MJ (2003) Research methods: managing primary study quality in meta-analyses. Res Nurs Health 26:322–333
Deeks JJ, Dinnes J, D'Amico R, Sowden AJ, Sakarovitch C, Song F, Petticrew M, Altman DG (2003) Evaluating non-randomised intervention studies. Health Technol Assess 7:1–173, iii-x
DerSimonian R, Laird N (1986) Meta-analysis in clinical trials. Control Clin Trials 7:177–188
Doi SA, Thalib L (2008) A quality-effects model for meta-analysis. Epidemiology 19:94–100
Doi SA, Thalib L (2009) An alternative quality adjustor for the quality effects model for meta-analysis. Epidemiology 20:314
Doi SA, Barendregt JJ, Mozurkewich EL (2011) Meta-analysis of heterogenous clinical trials: an empirical example. Contemp Clin Trials 32:288–298
Doi SA, Barendregt JJ, Onitilo AA (2012) Methods for the bias adjustment of meta-analyses of published observational studies. J Eval Clin Pract. doi:10.1111/j.1365-2753.2012.01890.x [Epub ahead of print]
Downs SH, Black N (1998) The feasibility of creating a checklist for the assessment of the methodological quality both of randomised and non-randomised studies of health care interventions. J Epidemiol Community Health 52:377–384
Egger M, Juni P, Bartlett C, Holenstein F, Sterne J (2003) How important are comprehensive literature searches and the assessment of trial quality in systematic reviews? Empirical study. Health Technol Assess 7:1–76
Eisenhart C (1947) The assumptions underlying the analysis of variance. Biometrics 3:1–21
Greenland S (1994) Invited commentary: a critical look at some popular meta-analytic methods. Am J Epidemiol 140:290–296
Herbison P, Hay-Smith J, Gillespie WJ (2006) Adjustment of meta-analyses on the basis of quality scores should be abandoned. J Clin Epidemiol 59:1249–1256
Higgins JP, Thompson SG (2002) Quantifying heterogeneity in a meta-analysis. Stat Med 21:1539–1558
Juni P, Witschi A, Bloch R, Egger M (1999) The hazards of scoring the quality of clinical trials for meta-analysis. JAMA 282:1054–1060
Kjaergard LL, Villumsen J, Gluud C (2001) Reported methodologic quality and discrepancies between large and small randomized trials in meta-analyses. Ann Intern Med 135:982–989

Leeflang M, Reitsma J, Scholten R, Rutjes A, Di Nisio M, Deeks J, Bossuyt P (2007) Impact of adjustment for quality on results of metaanalyses of diagnostic accuracy. Clin Chem 53:164–172

Lindsey JK (1999) On the use of corrections for overdispersion. Appl Stat 48:553–561

McCullagh P, Nelder JA (1983) Generalized linear models. Chapman and Hall, London

Moher D, Pham B, Jones A, Cook DJ, Jadad AR, Moher M, Tugwell P, Klassen TP (1998) Does quality of reports of randomised trials affect estimates of intervention efficacy reported in meta-analyses? Lancet 352:609–613

Moja LP, Telaro E, D'Amico R, Moschetti I, Coe L, Liberati A (2005) Assessment of methodological quality of primary studies by systematic reviews: results of the metaquality cross sectional study. BMJ 330:1053

Overton RC (1998) A comparison of fixed-effects and mixed (random-effects) models for meta-analysis tests of moderator variable effects. Psychol Methods 3:354–379

Poole C, Greenland S (1999) Random-effects meta-analyses are not always conservative. Am J Epidemiol 150:469–475

Realini JP, Goldzieher JW (1985) Oral contraceptives and cardiovascular disease: a critique of the epidemiologic studies. Am J Obstet Gynecol 152:729–798

Schulz KF, Chalmers I, Hayes RJ, Altman DG (1995) Empirical evidence of bias. Dimensions of methodological quality associated with estimates of treatment effects in controlled trials. JAMA 273:408–412

Senn S (2007) Trying to be precise about vagueness. Stat Med 26:1417–1430

Shuster JJ (2010) Empirical vs natural weighting in random effects meta-analysis. Stat Med 29:1259–1265

Slim K, Nini E, Forestier D, Kwiatkowski F, Panis Y, Chipponi J (2003) Methodological index for non-randomized studies (minors): development and validation of a new instrument. ANZ J Surg 73:712–716

Spiegelhalter DJ, Best NG (2003) Bayesian approaches to multiple sources of evidence and uncertainty in complex cost-effectiveness modelling. Stat Med 22:3687–3709

Tjur T (1998) Nonlinear regression, quasi likelihood, and overdispersion in generalized linear models. Am Stat 52:222–227

Tritchler D (1999) Modelling study quality in meta-analysis. Stat Med 18:2135–2145

Turner RM, Spiegelhalter DJ, Smith GC, Thompson SG (2009) Bias modelling in evidence synthesis. J R Stat Soc Ser A Stat Soc 172:21–47

Verhagen AP, de Vet HC, de Bie RA, Kessels AG, Boers M, Bouter LM, Knipschild PG (1998) The Delphi list: a criteria list for quality assessment of randomized clinical trials for conducting systematic reviews developed by Delphi consensus. J Clin Epidemiol 51:1235–1241

Verhagen AP, de Vet HC, de Bie RA, Boers M, van den Brandt PA (2001) The art of quality assessment of RCTs included in systematic reviews. J Clin Epidemiol 54:651–654

Wells G, Shea B, O'Connell D, Peterson J, Welch V, Losos M, Tugwell P (2000) The Newcastle-Ottawa Scale (NOS) for assessing the quality of nonrandomised studies in meta-analyses. http://www.ohri.ca/programs/clinical_epidemiology/oxford.htm. Accessed 15 June 2007

Whiting P, Harbord R, Kleijnen J (2005) No role for quality scores in systematic reviews of diagnostic accuracy studies. BMC Med Res Methodol 5:19

Woolf B (1955) On estimating the relation between blood group and disease. Ann Hum Genet 19:251–253

Chapter 15
Meta-analysis II

Interpretation and Use of Outputs

Adedayo A. Onitilo, Suhail A.R. Doi, and Jan J. Barendregt

Abstract Outputs in meta-analysis give us measures of evidence dissemination bias or graphical representation of the pooled results and their underlying heterogeneity. This chapter discusses the various outputs with a focus on their utility and interpretation. Examples focus on the use of MetaXL, which is our own software developed for meta-analysis and is freely available from www.epigear.com. This is the only software currently available that can perform a bias-adjusted meta-analysis.

Introduction

The main output of a meta-analysis is the pooled estimate and its confidence interval. In addition, there are also a number of graphical and numerical outputs that aid with interpretation of results by presenting information such as study heterogeneity, detection of publication bias, and other important aspects of the meta-analysis. Graphical and statistical representation should not replace, but should be used in addition to narrative description of study design, setting, methods, follow-up analysis methods as well as the strengths and limitations of the individual studies pooled together in the meta-analysis. MetaXL is our

A.A. Onitilo
Marshfield Clinic - Weston Center, Weston, USA

University of Queensland, School of Population Health, Brisbane, QLD, Australia

University of Wisconsin, School of Medicine and Public Health, Madison, USA

S.A.R. Doi (✉)
University of Queensland, School of Population Health, Brisbane, QLD, Australia

Princess Alexandra Hospital, Brisbane, Australia
e-mail: sardoi@gmx.net

J.J. Barendregt
University of Queensland, School of Population Health, Brisbane, QLD, Australia

S.A.R. Doi and G.M. Williams (eds.), *Methods of Clinical Epidemiology*,
Springer Series on Epidemiology and Public Health,
DOI 10.1007/978-3-642-37131-8_15,

preferred software for meta-analysis (downloadable freely from www.epigear.com) and all outputs we discuss use MetaXL as far as possible. In addition, bias-adjusted meta-analyses can only be run using MetaXL.

Individual and Pooled Results

Forest Plot

The forest plot, a graphical presentation of meta-analysis results, first used in 1982 by Lewis and Ellis, has now become standard practice and is arguably the most important output from a meta-analysis. A forest plot presents individual study estimates, the pooled estimate, confidence intervals, as well as the weight of each study in the analysis and heterogeneity statistics. Individual studies used in the meta-analysis are represented in the plot by horizontal lines; the length of each line represents the confidence interval around each study estimate. Shorter lines represent a narrower confidence interval thus higher precision of study effect size (ES), usually found in larger studies. Conversely, longer lines represent a wider confidence interval and less precision around effect side, usually found in smaller studies. The point estimate from each study is represented by a shape on the line such as a dot or box, and the size of this shape represents the weight of the study in the meta-analysis. A summary estimate of the point estimates is also represented by a shape at the end of the graph. Most forest plots also have two vertical lines. A dotted vertical line represents the pooled estimate and a solid vertical line represents the null estimate. For example, for the odds ratio this is 1 and for the mean difference the null has a value of 0. Horizontal lines that cross the null vertical line represent non-significant studies. Most forest plots in meta-analysis will arrange studies in chronologic order or by subgroups. This allows for further subgroup analysis or stratification. It can also be a way to represent heterogeneity in the meta-analysis. The plot can either be on a normal scale or logarithmic scale. The normal scale is usually used for mean difference and rates, while logarithm scales are used for ratios.

Figure 15.1 presents the forest plot from a quality effects model analysis of patient mortality before and after changes to the working hour regulations for surgeons. The square on the plot for individual studies is proportional to the weight it has in the meta-analysis; the horizontal lines represent the study's confidence interval. The dotted vertical line on the right gives the pooled estimate, the solid vertical line on the left is the null result, in this case $OR = 0$. Inspection of the forest plot can give a good indication of the amount of heterogeneity. In MetaXL, the forest plot is obtained by choosing Results from the MetaXL menu.

Forest plots are easy to read and interpret, although one drawback is that attention is often drawn to the least precise study which has the longest horizontal line and actually carries less weight in the meta-analysis. While the forest plots

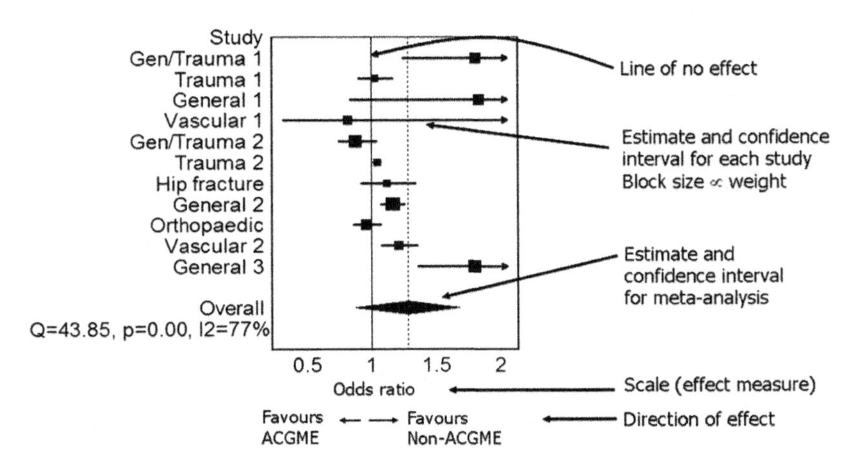

Fig. 15.1 The forest plot from a quality effects model analysis of mortality after (compared with before) the ACGME (Accreditation Council for Graduate Medical Education) regulations that reduced working hours for surgeons. The *dotted vertical line* indicates the pooled effect size. Q is the Cochran Q statistic for heterogeneity and I2 is the I^2 statistic and both are discussed below

should be considered whenever feasible and appropriate in reporting a meta-analysis it may not be the most appropriate representation of meta-analysis that involves too many studies, in this case a summary forest plot or other plots such as the Galbraith plot should be considered. In the summary forest plot, individual study results are replaced with pooled results from either different outcomes or different subgroups. Thus individual points represent meta-analyses rather than studies.

Sensitivity Analysis

Sensitivity analysis explores the ways in which the main findings are changed by varying the selection criteria for studies that are combined. The sensitivity analysis is executed by running the meta-analysis across categories of selected studies; for example, published versus unpublished studies or other selection criteria based on patient group, type of intervention or setting. A meta-analysis can also be performed by leaving out one study at a time to see if any single study has a large influence on the pooled results. Sensitivity analysis can also be done by running different meta-analysis models to examine the robustness of the method utilized in the meta-analysis. Usually if there is no significant heterogeneity in the studies used, most methods should yield comparable summary estimates. When dose–response or open-ended variables are examined in the meta-analysis, a sensitivity analysis can limit the range in the dose–response or open-ended variable that produce most of the effect. In meta-analyses without sensitivity analyses, the likely

Table 15.1 Leave-one-out sensitivity analysis results (Charlson et al. 2012)

	Meta-analysis results with study excluded			Weight (%) of the study in the complete analysis
Study excluded	ES	Lower 95 % confidence limit	Higher 95 % confidence limit	
Gangadhar	−0.897	−1.540	−0.255	18.8
Vieweg	−0.190	−0.907	0.527	14.2
Lerer and Segman	−0.138	−0.846	0.570	17.8
Shapira	0.497	−0.084	1.078	25.5
Kellner	0.892	−0.027	1.811	8.0
Janakiramaiha (L)	1.169	0.220	2.117	9.6
Janakiramaiha (H)	2.236	0.944	3.528	6.0

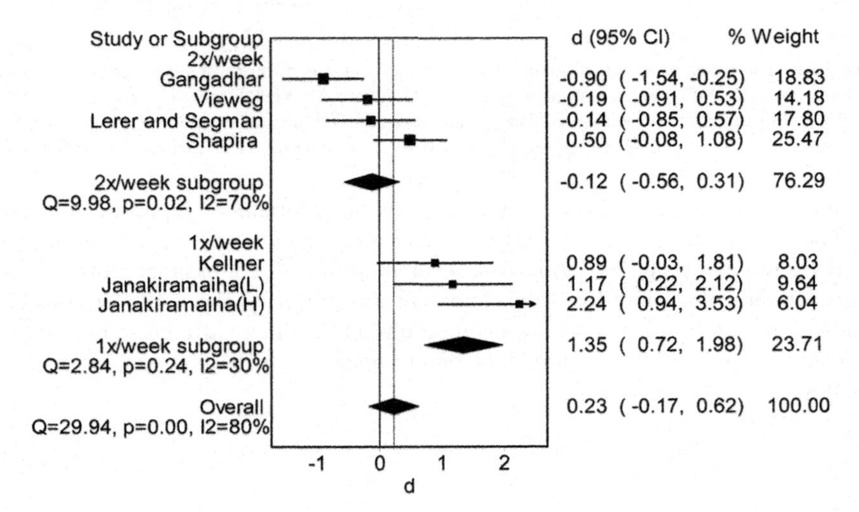

Fig. 15.2 Forest plot comparing 3×/week ECT with 2×/week (*upper* subgroup) or 1×/week (*lower* subgroup). It is evident that the 1×/week frequency results in a greater difference from 3×/week compared with 2×/week (Data from Charlson et al. 2012). Heterogeneity is diminished within subgroups suggesting that ECT frequency contributes to overall heterogeneity

impact of these important factors on the key finding is ignored and thus the results are less robust. An example of a leave-one-out sensitivity analysis is given in Table 15.1 and also depicted in the forest plot for 2×/week versus 1×/week comparisons after electroconvulsive therapy (ECT) for depression (Fig. 15.2).

Heterogeneity

One of the most important aspects of meta-analysis is to determine whether heterogeneity exists in the studies combined in the analysis and investigate the source of such heterogeneity. It underscores the use of meta-analysis as a means of generating a summary estimate, lends conclusiveness to otherwise inconclusive clinical situations and extends meta-analysis to explain differences between the combined studies. Heterogeneity can be clinical or statistical. Clinical heterogeneity is based on the characteristics of studies combined (e.g., study design, follow-up length, duration of therapy) and characteristics of subjects in the studies. Thus clinical heterogeneity can be within or between studies. Statistical heterogeneity refers to situations when the estimates from different studies deviate considerably from each other. Below we describe some of the formal statistical tests and plots for assessing heterogeneity.

Cochran's Q

Cochran's Q is a heterogeneity statistic. It is the classical measure of heterogeneity and is given by

$$Q = \sum_i w_i (\theta_i - \theta_\mathrm{p})^2$$

where i is an index for the study, w_i is the fixed effect weight of study i, θ_i is the estimate from study i, and θ_p is the fixed effect pooled estimate. The Q statistic follows a chi-squared distribution with $k - 1$ degrees of freedom under the null hypothesis of homogeneity, where k is the number of studies in the meta-analysis. If the probability of the value of Q occurring by chance is low ($p < 0.05$), the null hypothesis is rejected and heterogeneity is assumed. Unfortunately, the Q statistic is not very sensitive when the number of studies is not large. In that case, some authors prefer a critical value for p of 0.1 instead of 0.05. When the number of studies is large, the Q statistic becomes too sensitive. In MetaXL the MACochranQ function returns the Q statistic in the spreadsheet. The test can then be performed using Excel's CHIDIST function. The Q statistic and its test result are also presented in the forest plot and tabular output.

The magnitude of the computed Q is dependent on the weight and the number of studies in the meta-analysis. If there are limited number of small studies (<20 studies), it has been shown that the asymptotic Q statistic gives the correct type I error under the null hypothesis but has low power (Takkouche et al. 1999) and null for heterogeneity is not likely to be rejected. Whereas if there are large number of studies or large sample size studies in the meta-analysis, irrespective of true clinical heterogeneity Q has too much power and null for heterogeneity is likely to be rejected. For this reason, it is always important to examine the studies in the meta-analysis for clinical heterogeneity.

I^2

The I^2 statistic is another means to detect heterogeneity, and is derived from the Q statistic. The I^2 examines the percentages of variation across studies due to heterogeneity rather than by chance and it is given by

$$I^2 = \begin{cases} 100\frac{Q-\mathrm{df}}{Q} & \text{if} > 0 \\ 0 & \text{otherwise} \end{cases}$$

where $\mathrm{df} = k - 1$ is degrees of freedom. Confidence intervals for I^2 can be derived as follows:

Define $H = \sqrt{Q/(k-1)}$. Then,

$$\mathrm{SE}[\ln(H)] = \begin{cases} \frac{1}{2}\frac{\ln(Q)-\ln(k-1)}{\sqrt{2Q}-\sqrt{2k-3}} & \text{if } Q > k \\ \sqrt{\left\{\frac{1}{2(k-2)}\left(1-\frac{1}{3(k-2)^2}\right)\right\}} & \text{otherwise} \end{cases}$$

95 % confidence intervals for H are then derived by

$$\exp(\ln H \pm 1.96\mathrm{SE}[\ln(H)])$$

Since

$$I^2 = \frac{H^2 - 1}{H^2}$$

the confidence intervals for I^2 are derived from those of H. The I^2 statistic is thus a number between 0 and 100. A rule of thumb is that heterogeneity is low for an I^2 of 25, moderate for an I^2 of 50, and high for an I^2 of 75. In MetaXL the MAISquare function returns the I^2 statistic in the spreadsheet. It is also presented in the forest plot and tabular output; the latter includes the confidence interval. Effectively, I^2 is $(Q - (k - 1))$ divided by Q where k denotes the number of studies. I^2 has the same problem of low statistical power with small numbers of studies. Specifically, the confidence intervals around I^2 behave very similarly to tests of Q in terms of type I error and statistical power. Also, I^2 increases with the number of subjects included in the studies in a meta-analysis. It thus seems counterintuitive to criticize Q as having low power on the one hand and to define a measure (and an assessment rule) that would require the heterogeneity test to be even more significant. From the point of view of validity, power and computational ease, the Q statistic is probably a better choice compared with I^2. Unlike the Q statistic, the I^2 statistic does not vary based on the number of studies included in the meta-analysis, it is possible to compare the statistical heterogeneity of meta-analyses with different numbers of studies. However, I^2 will tend to increase artificially as evidence accumulates since it increases with number of subjects included in the meta-analysis. Additionally, as I^2 is the percentage of variability that is due to between-study heterogeneity, $1 - I^2$ is the percentage of variability that is due to sampling error. When the studies become very large, the sampling error tends to 0 and I^2 tends to 1 (Rucker et al. 2008). Such heterogeneity may not be clinically relevant and studies with relatively large I^2 in this situation may still be usefully pooled if other measures such as τ^2 remain relatively small and clinically relevant heterogeneity is unlikely to be present.

τ^2

Yet another statistic is τ^2, which is the random effects variance component calculated as part of a random effects meta-analysis. The τ^2 statistic examines the between-study variance and is given by

$$\tau^2 = \frac{Q - (k - 1)}{\sum w_i - \left(\frac{\sum w_i^2}{\sum w_i}\right)}$$

which is set to zero if $Q < k - 1$, and w_i is the inverse variance weight. The τ^2 statistic is the variance of the presumed normally distributed individual study estimates under the assumptions of the random effects model.

In MetaXL the MATauSquare function returns the τ^2 statistic in the spreadsheet, and it is also presented in the tabular output of random effects analyses. It may also be used as a marker of heterogeneity if its value is greater than zero. Similar to the Q and I^2 statistic the τ^2 statistic has its limitation; it is not very powerful if the number of studies is small or if the conditional variances between the studies are large. The advantage, however, is that it does not depend on the number or size of studies in the meta-analysis, i.e., it can be kept fixed with increasing subjects in the meta-analysis. Furthermore, since τ^2 is measured on the same scale as the outcome, it can therefore be directly used to quantify variability. Note that assessment of τ^2 does not give us a p value but rather a yes/no answer only, and certainly there will be little heterogeneity if $\tau^2 = 0$ regardless of the value of I^2. We must keep in mind however that τ^2 assumes normality of the random effects and the error terms.

Q Index

The Q index is applicable to the quality effects model only. It expresses the percentage of study weight that is re-distributed in the quality effects analysis. It is given by

$$Q_{\text{index}} = 100\left(\sum_i w_i \frac{(1-q_i)}{\sum_j w_j}\right)$$

where q_i is the quality score of study i and w_i is the inverse variance weight.

In MetaXL the MAQIndex function returns the Q index statistic in the spreadsheet, and it is also presented in the tabular output of quality effects analyses. The Q index is the only measure that inputs study quality as a source of heterogeneity. It therefore has the advantage of imputing clinical heterogeneity in statistical terms, a strength not seen in any other statistical test for heterogeneity.

Galbraith Plot

The Galbraith plot (Fig. 15.3) presents standardized effect estimate on the vertical axis plotted against the inverse of the standard error on the horizontal axis. It is a linear regression constrained through the origin of the standardized treatment effects (treatment effect divided by its standard error) on their inverse standard errors which yields a regression line. Typically a dotted line is used at ±2SD confidence interval above and below this line. The slope of the regression line provides details of the unstandardized effect estimates. Galbraith plots facilitate examination of heterogeneity, including detection of outliers. With a fixed effect model, 95 % of studies in a meta-analysis should be found on this plot to be within the two confidence interval lines and the more precise studies are farthest from the origin of the linear regression line. Different symbols can be used in the plots to represent sub-sets or stratification thus making identification of the source of heterogeneity easier. Also the graph can be labelled to show the direction the effect the estimate favors. Compared to the forest plot, the Galbraith plot is able to display more studies that cannot be easily done by the forest plot and it also has the additional advantage that it gives a better representation of heterogeneity.

L'Abbé Plot

The L'Abbé plot is used to present the results of multiple clinical trials with dichotomous outcomes showing for each study; the observed event rate in the experimental group plotted against the observed event rates in the control group. It is used to view the range of event rates among the trials and highlight excessive heterogeneity. The L'Abbé plot is also ideally suited to diagnostic meta-analyses (Fig. 15.4) where diseased (group 1) and healthy (group 2) rates of test positivity can be compared across studies. The shape representing each study is usually proportional to the size of each study (or study weights) since unlike the forest

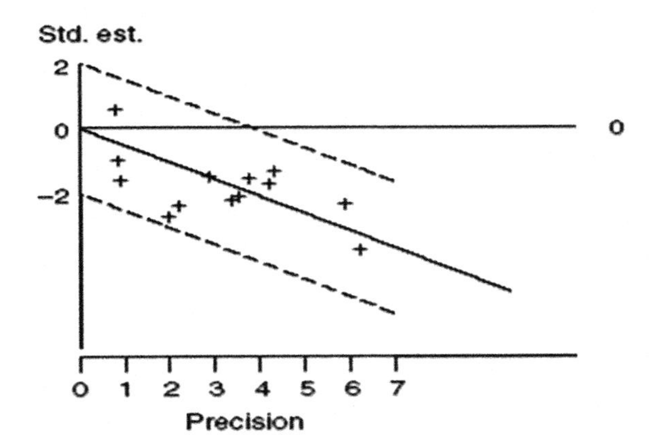

Fig. 15.3 A Galbraith Plot with standardized effect estimate on the vertical axis plotted against the inverse of the standard error as a measure of precision on the horizontal axis. The intercept is constrained to zero. *Solid lines* represent the unweighted regression line constrained at 0 with a slope equal to the overall effect size of a fixed effects meta-analysis, and its 95 % confidence intervals (*dashed line*). The position of the studies on the y-axis indicates their contribution to the Q statistic for heterogeneity. The position of the studies on the x-axis indicates the weight of each study in the meta-analysis

plot or Galbraith plot there is no information about the precision of the studies on the plotted axes. The L'Abbé plot should be considered when outcomes are dichotomous across studies (treatment vs. control) or for diagnostic studies (sensitivity vs. false-positive rates).

Publication Bias

Publication bias refers to the phenomenon whereby studies with significant outcomes are more likely to be submitted for publication compared to null result or non-significant studies. This is usually assessed by several statistical and graphical (quasi-statistical) means.

Funnel Plot

Funnel plots assess publication bias or heterogeneity by plotting the trials' effect estimates against a measure of precision, Asymmetrical plots are interpreted to suggest that selection biases are present. The use of such a plot is based on the fact that precision in estimating the underlying treatment effect will increase as the precision of the study increases and thus results from small studies will scatter widely at the bottom of the plot, with the spread narrowing with increasing

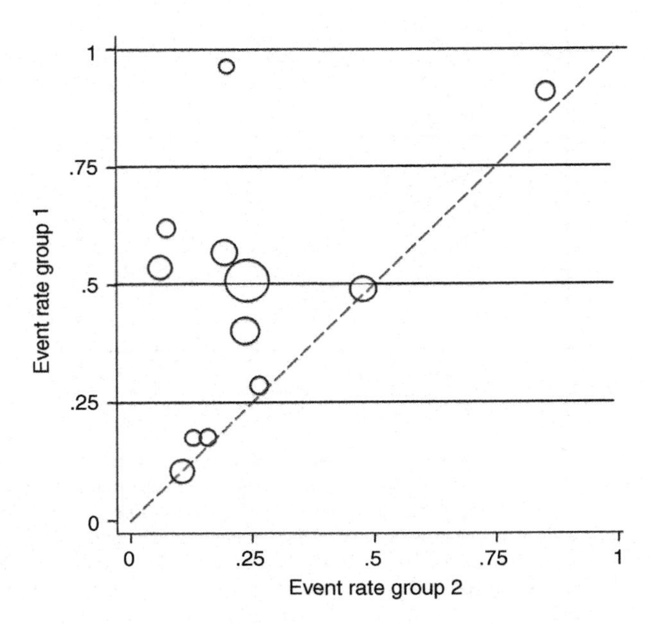

Fig. 15.4 L'Abbé plot demonstrating true-positive (group 1) and false-positive (group 2) rates in diseased and healthy subjects, respectively, in a diagnostic meta-analysis (Data used from Whiting et al. 2005)

precision. In the absence of selection bias, the plot is expected to resemble a symmetrical inverted funnel. It usually recommended that ratio measures of intervention effect should be plotted on a logarithmic scale, so that effects of the same magnitude but opposite directions (e.g., odds ratios of 0.5 and 2) are equidistant from 1.0.

Figure 15.5 shows the funnel plot from the meta-analysis of fibrinolysis in myocardial infarction study. The vertical line represents the pooled estimate from the inverse variance model; the funnel sides represent the 95 % confidence intervals around the pooled estimate, given the standard error on the *y*-axis; and the dots represent the individual study results. The aim of the funnel plot is to examine publication bias. When the study dots are largely symmetrical around the pooled estimate, there is no evidence for publication bias. In the present case, there is a large degree of asymmetry, which suggests publication bias is present. Funnel plots can look quite different, depending on the choice of *y*-axis. MetaXL offers three options: inverted standard error (default, and used in Fig. 15.5), precision, and inverse variance. For the log of risk or odds ratio, the inverted standard error is recommended. In MetaXL the funnel plot is obtained by choosing Results from the MetaXL menu.

While there has been much focus on selection biases in relation to the association between size and effect in a meta-analysis, it must be kept in mind that asymmetry can also occur for reasons other than selection biases due to selective

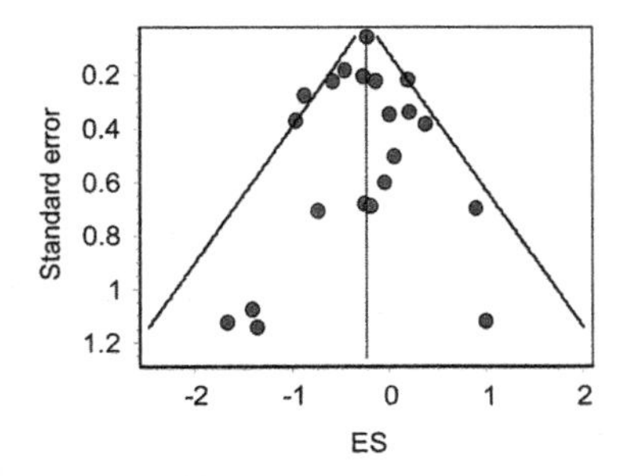

Fig. 15.5 The funnel plot from the fibrinolysis in myocardial infarction meta-analysis (Yusuf et al. 1985). The ES is the LnOR. The *central line* depicts the fixed effects pooled estimate and the limbs of the funnel are made up by the limits of the confidence interval around the pooled estimate being computed successively based on the standard error of each study

publication or selective outcome reporting. These other factors related to study size include study quality (smaller usually thought to be worse), presence of true heterogeneity (e.g. different baseline risks in small and large studies), an association between the intervention effect and its standard error (artefactual) or even chance. Despite the initial expectations, assessment of publication bias using the classic funnel plot continues to misrepresent bias because the appearance of the standard funnel plot has been shown to be misleading. Furthermore, it has been demonstrated that discrepancies between large trials and corresponding meta-analyses and heterogeneity in meta-analyses may also be largely dependent on the arbitrary choice of the method used to construct the classic funnel plot. In particular, the shape of the plot in the absence of bias changes with the choice of axes and it has been suggested that funnel plots of meta-analyses should generally be limited to using standard error as the measure of study size and ratio measures of treatment effect. Even when this is adhered to, the visual and quantitative assessment of asymmetry is flawed. It has been suggested that funnel plot asymmetry detected using measures of impact such as the risk difference (measures that are correlated with baseline risk) may be artefactual and thus funnel plots and related tests using risk differences should not be undertaken.

Egger's Regression

The most popular formal statistical test of funnel plot asymmetry is the Egger's test. Its power is limited, particularly for moderate amounts of bias or meta-analyses based on a small number of small studies. Egger's regression is essentially a linear regression on the standardized ESs (Zi) with precision $(1/\sigma_i^2)$as predictor where θ_i is the ES and the standardized ES is then given by

$$z_i = \frac{\theta_i}{\sigma_i}$$

Egger's regression is then

$$z_i = \alpha + \beta \frac{1}{\sigma_i}$$

With no publication bias present, the intercept (α) should not be significantly different from zero. This is similar to regression of a Galbraith plot not constrained to the origin (see above).

Doi Plot

Another plot that is more objective uses the approach of a linear ranking to assess study asymmetry using the same scale for the ES on which its standard error exists. Essentially, each subject in every trial within the meta-analysis is assigned the ES of their trial and ranked serially. As all subjects in a trial have the same ES, they will have the same rank and thus each trial has a single final rank based on the number of subjects (*N*) in the study. However, because *N* does not capture the trials' information content completely (the number of observed events in each arm of a study is often more important in driving the precision of the estimate than the study size per se), an updated *N* (designated N') is used to incorporate this. The final ranking is then converted to a percentile and then a *z*-score using the method detailed below.

First, N' is generated as follows:

$$N'_i = \text{int}\left(N_i \times \frac{\left(\max\{\text{SE}_i{}^2 N_i\}\right)}{\text{SE}_i{}^2 N_i}\right)$$

where SE is the standard error of the ES. If there are *k* studies in a meta-analysis numbered serially as $i = 1,\ldots,k$ each with an ES and study-adjusted patient-information study size (N'), the *k* studies can then be ranked by ES and the N' subjects in these *k* trials are serially numbered consecutively. The last subject number in each study (A_i) is determined by summing the N'_i across trials with ES smaller than or equal to the ES under consideration then (using indicator functions):

$$A_i = \sum \left\{ \left(N'_1 \times 1_{\{\text{ES}_1 \leq \text{ES}_i\}}\right), \ldots, \left(N'_k \times 1_{\{\text{ES}_k \leq \text{ES}_i\}}\right) \right\}$$

If we assign all subjects in a trial to the ES of their trial, the final rank (R_i) of each study based on ES and number of subjects is computed as follows:

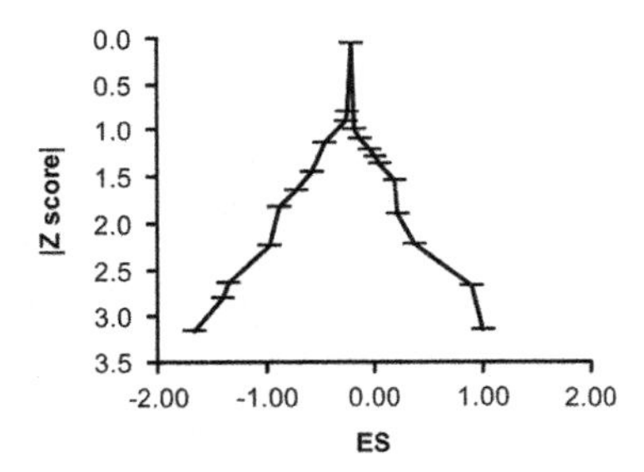

Fig. 15.6 The Doi plot for the same studies as in Fig. 15.5. The ES is the LnOR. The *dashes* represent each study with the absolute Z score on the *y*-axis plotted against the effect size on the *x*-axis. Absence of publication bias is indicated by a similar slope of each limb of the plot away from the vertical plane. With publication bias, both limbs will slope differently with respect to the vertical plane

$$R_i = \frac{\max\{(A_1 \times 1_{\{ES_1 < ES_i\}}), \ldots, (A_k \times 1_{\{ES_k < ES_i\}})\} + A_i}{2}$$

R_i is then converted into a percentile (P_i) as follows:

$$P_i = \frac{(R_i - 0.5)}{\sum_{i=1}^{k} N_i}$$

Finally the percentile is converted into a *z*-score [$z = \text{NORMINV}(P_i,0,1)$].

This new measure of precision is now the absolute value of the *z*-score and the ES is then plotted against this absolute value of the *z*-score to create the new mountain plot. With symmetrical studies, the most precise trials will define the mid-point around which results should scatter, and thus they will be close to mid-rank and will be close to zero on the *z*-score axis. Smaller less precise trials will produce an ES that scatters increasingly widely, and the absolute *z*-score will gradually increase for both smaller and larger ES's on either side of that of the precise trials. Thus, a symmetrical triangle is created with a *z*-score close to zero at its peak. If the trials are homogeneous and not affected by selection or other forms of bias, the plot will therefore resemble a symmetrical triangle with the studies themselves making up the limbs of the plot (Fig. 15.6).

Bibliography

Bax L, Ikeda N, Fukui N, Yaju Y, Tsuruta H, Moons KG (2009) More than numbers: the power of graphs in meta-analysis. Am J Epidemiol 169:249–255

Charlson F, Siskind D, Doi SA, McCallum E, Broome A, Lie DC (2012) ECT efficacy and treatment course: a systematic review and meta-analysis of twice vs thrice weekly schedules. J Affect Disord 138:1–8

Lewis JA, Ellis SH (1982) A statistical appraisal of post-infarction beta-blocker trials. Primary Cardiol suppl 1:31–37

Rücker G, Schwarzer G, Carpenter JR, Schumacher M (2008) Undue reliance on I2 in assessing heterogeneity may mislead. BMC Med Res Method 8:79

Rucker G, Carpenter JR, Schwarzer G (2011) Detecting and adjusting for small-study effects in meta-analysis. Biom J 53:351–368

Sterne JA, Egger M (2001) Funnel plots for detecting bias in meta-analysis: guidelines on choice of axis. J Clin Epidemiol 54:1046–1055

Sterne JA, Sutton AJ, Ioannidis JP, Terrin N, Jones DR, Lau J, Carpenter J, Rucker G, Harbord RM, Schmid CH, Tetzlaff J, Deeks JJ, Peters J, Macaskill P, Schwarzer G, Duval S, Altman DG, Moher D, Higgins JP (2011) Recommendations for examining and interpreting funnel plot asymmetry in meta-analyses of randomised controlled trials. BMJ 343:d4002

Takkouche B, Cadarso-Suarez C, Spiegelman D (1999) Evaluation of old and new tests of heterogeneity in epidemiologic meta-analysis. Am J Epidemiol 150:206–215

Tang JL, Liu JL (2000) Misleading funnel plot for detection of bias in meta-analysis. J Clin Epidemiol 53:477–484

Terrin N, Schmid CH, Lau J (2005) In an empirical evaluation of the funnel plot, researchers could not visually identify publication bias. J Clin Epidemiol 58:894–901

Whiting P, Harbord R, Kleijnen J (2005) No role for quality scores in systematic reviews of diagnostic accuracy studies. BMC Med Res Methodol 5:19

Yusuf S, Collins R, Peto R, Furberg C, Stampfer MJ, Goldhaber SZ, Hennekens CH (1985) Intravenous and intracoronary fibrinolytic therapy in acute myocardial infarction: overview of results on mortality, reinfarction and side-effects from 33 randomized controlled trials. Eur Heart J 6:556–585

Appendix: Stata Codes

Chapter 1

Table 1.4

```
kap rater1A rater2A, tab
kap rater1B rater2B, tab
kap rater1C rater2C, tab
kap rater1D rater2D, tab
```

Categorical Scale Example

```
kap rater1 rater2, tab
```

Table 1.5

```
kap rater1 rater2, tab wgt(w)
kap rater1 rater2, tab wgt(w2)
```

(note that w implies linear weight and w2 implies quadratic weight)

S.A.R. Doi and G.M. Williams (eds.), *Methods of Clinical Epidemiology*,
Springer Series on Epidemiology and Public Health,
DOI 10.1007/978-3-642-37131-8,

Chapter 10

Table 10.1

```
bysort gender: ci death_yr, binomial wald
tab gender death_yr, chi2 nofreq
bysort age_group: ci death_yr, binomial wald
tab age_group death_yr, chi2 nofreq
```

Table 10.2

```
cs death_yr gender, or
cs death_yr age_group if age_group==0 | age_group==1, or
cs death_yr age_group if age_group==0 | age_group==2, or
cs death_yr age_group if age_group==0 | age_group==3, or
```

Table 10.3

```
foreach var of varlist sho chf mitype cvd afb av3 miord{
cs death_yr `var', or
}
```

(Please note that the second line syntax is `var' not 'var')

Table 10.4

```
recode age (min/69 = 0) (70/max = 1), gen(age_70)
cs death_yr sho if age_70==0, or woolf
cs death_yr sho if age_70==1, or woolf
```

Table 10.5

```
cs death_yr sho, by(age_70)
```

Table 10.6

```
xi: logit death_yr i.age_group
xi: logit death_yr i.age_group, or
```

Table 10.7

```
xi: logit death_yr i.age_group sho
xi: logit death_yr i.age_group sho, or
```

Table 10.8

```
xi: logit death_yr i.age_group sho
estimates store age_shock
xi: logit death_yr i.age_group
estimates store age
lrtest age_shock age
xi: logit death_yr sho
estimates store shock
lrtest age_shock shock
```

Table 10.9

```
logit death_yr age
logit death_yr age, or
```

Table 10.10

```
logit death_yr age
sort age
preserve
drop if age[_n] == age[_n-1]
gen predicted_index = 0
gen predicted_prob = 0
gen odds = 0
forvalues age = 55/84 {
```

```
replace predicted_index = _b[_cons] + (_b[age]*age)
replace predicted_prob = exp(predicted_index) / (1+exp
(predicted_index))
}
forvalues age = 55/84 {
replace odds = predicted_prob/(1-predicted_prob)
}
gen oddsratio = odds[_n]/odds[_n-1]
gen relativerisk = predicted_prob[_n]/predicted_prob
[_n-1]
list age predicted_prob oddsratio relativerisk if
age>=55 & age<=84
restore
```

Table 10.11

```
generate age_centred = age - 70
generate age_quadratic = (age - 70)^2
logit death_yr age_centred age_quadratic
logit death_yr age_centred age_quadratic, or
```

Table 10.12

```
logit death_yr age 0.gender sho, or
```

Table 10.13

```
logit death_yr age gender sho
estimates store age_gender_shock
logit death_yr age gender
estimates store age_gender
logit death_yr age  sho
estimates store age_shock
logit death_yr  gender sho
estimates store gender_shock
lrtest age_gender_shock age_gender
lrtest age_gender_shock age_shock
lrtest age_gender_shock gender_shock
```

Table 10.14

```
cs death_yr chf if age_70==0, or
cs death_yr chf if age_70==1, or
```

Table 10.15

```
xi: logit death_yr i.age_70 i.chf
xi: logit death_yr i.age_70 i.chf, or
```

Table 10.16

```
logit death_yr i.age_70##i.chf
logit death_yr i.age_70##i.chf, or
```

Table 10.17

```
logit death yr i.age 70#i.chf i.age 70, or
```

Table 10.18

```
logit death_yr i.sho i.chf#i.age_70 i.age_70
logit death_yr i.sho i.chf#i.age_70 i.age_70, or
```

Chapter 11

Table 11.1

```
list id time censor tx strat2 in 1/20
```

Table 11.2

```
ltable time censor if tx==0
```

Table 11.3

```
stset time, failure(censor==1)
sts list if tx==0, at(0/364)
sts graph if tx==0, tmax(30)lost
```

Figure 11.2

```
sts graph if tx==0, ci risktable ytitle(Survival) xtitle
(Time (days)) legend(off) title(Survival curve for two-
drug therapy)
```

Figure 11.3

```
sts graph if tx==0, failure ci risktable yscale(range
(0 0.2)) ylabel(0(0.05)0.2) ytitle(Cumulative incidence)
xtitle(Time (days)) legend(off) title(Failure curve for
two-drug therapy)
```

Figure 11.4

```
sts graph, by(tx) ci yscale(range(0.8 1)) ylabel(0.8(0.5)1)
ytitle(Survival) xtitle(Time (days)) title(Survival curve
for both therapies)
```

Table 11.4

```
sts test tx
```

Figure 11.6

```
stset lenfol, failure(fstat==1)
sts graph, ci risktable ytitle(Survival) xtitle(Time
(days)) title(Survival curve) tmax(2000)
```

Figure 11.7

```
sts graph, by(chf) ci risktable ytitle(Survival)
xtitle(Time (days)) title(Survival curves comparing
CHF Complications) tmax(2000)
```

Table 11.5

```
stcox i.chf, nohr
stcox i.chf
```

Table 11.6

```
stcox i.gender, nohr
stcox i.gender
```

Table 11.7

```
recode age (min/64=1) (65/79=2) (80/max=3), gen
(age_category)
stcox i.age_category
stcox age
```

Table 11.8

```
stcox i.chf age i.gender
```

Table 11.9

```
tabstat age, by(gender) stat(n mean min max)
tab chf gender, col
```

Table 11.10

```
stset lenfol, failure(fstat==1)id(id)
stsplit period, at(1) after(_t=((2.5*365.25)-1))
stcox i.chf##i.period
```

Table 11.11

```
stcox i.chf#i.period i.period
```

Table 11.12

```
stcox i.chf##i.period age
```

Table 11.13

```
stcox i.chf##i.period i.period age
```

Index

S.A.R. Doi and G.M. Williams (eds.), *Methods of Clinical Epidemiology*, Springer Series on Epidemiology and Public Health,
DOI 10.1007/978-3-642-37131-8,

Printed by Printforce, the Netherlands